# T R A I L S

## *of a Wilderness Wanderer*

# *T R A I L S*

## *of a Wilderness Wanderer*

### ANDY RUSSELL

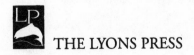

THE LYONS PRESS

*To my mother and father*

Printed in Canada

10  9  8  7  6  5  4  3  2  1

Library of Congress Cataloging-in-Publication Data

Russell, Andy, 1915–
Trails of a wilderness wanderer/Andy Russell.—Lyons Press reprint ed.
p.  cm.
Originally published: New York: Knopf, 1971. With introd.
ISBN 1-58574-183-3
1. Ranch life—Alberta. 2. Outdoor life—Alberta. 3. Frontier and pioneer
life—Alberta. 4. Trapping—Alberta. 5. Alberta—Social life and customs.
6. Russell, Andy, 1915–  7. Alberta—Biography. I. Title.

F1076.R9 2000

971.23'03—dc21

00-60835

Published by arrangement with Alfred A. Knopf, Inc.
Lyons Press reprint edition, 2000

# CONTENTS

# *INTRODUCTION*

## *to the New Edition*

WHEN I LOOK BACK OVER THE LONG TRAIL TO THE FAR horizon behind me, I am amazed at the pace of the world in which I've lived: from a primitive world of nature to the modern, high speed computerized tempo of today. Once I lived from hunting and gathering what was needed to stay alive in the wilds; now I make my living largely at a desk, trying to preserve the vital habitat required for the survival of fish and wildlife, as well as the continuance of Man.

It is very pleasant to look back to a time when the creeks and rivers pouring down from the rooftree of the continent were as clear, cold, and pure as freshly shattered glacier ice. We worked with cattle and horses, hunted, trapped and wandered with the satisfying feeling of being part of it all. We did not fight it. The wise ones lived content to be warm with a full belly under a roof that didn't leak on a stormy night. We had the smell of saddle leather, sweat, gunpowder and dust in our noses, and knew the vibrant thrill of a good fishing rod with the fight of trout on the end of the line. We knew the joy of being truly alive in the clear open air under a vast sky where the clouds ran in the wind like wild colts at play. We adventured, meeting things as they came without plan, reveling in the feel of a bucking horse on a cool morning or throwing a rope to catch a cow. Sometimes out of sheer exuberance and fun, we deliberately spooked something just to see how high it would go and where it would come down. It is one thing just to live, but something else to be really alive.

The smoke of our evening fires blended with stories of things we had seen and done, sometimes punctuated with music and song. For weeks we might not see an up-to-date newspaper, radios were something of a rarity, and television had not been invented. So we told stories. It was a contribution expected of everyone and our social status depended on how good we were at it.

I owe much to those oldtimers with whom I shared the evening fires, and told of many things not just to entertain but to instruct. They were generous to the kid who would tag along to watch what they were doing. My father, who had ridden on the round-ups on the prairies before there were any barbwire fences, was one of them. They taught my brother and me the skills of riding, roping, handling guns safely and shooting them accurately; it was the kind of education we had to have to live through some of the mix-ups that we sometimes found ourselves in.

They were the pioneers, and though some of them were real rogues, they were the trail blazers and empire builders who helped to open this big and wonderful country without hurting it very much on the way. We owe them much. Except for a few of us, who were the kids they taught, they are all gone.

Like Aldo Leopold, I am glad I am not a boy again with no wild place to grow up in. Some of the sons and grandsons of those oldtimers assume they can ignore the rules of nature by which their forbearers lived. They have yet to learn that if they throw nature out the door, it will only come back through the window.

We grew up knowing how to meet danger, cope with it and survive. The dangers we meet now are not just of the moment; they theaten us all. If we can hope to endure, we must deal with them as a cultural and global human entity, seeing beyond our individual, local horizons with renewed recognition of our place in the world of nature.

# INTRODUCTION

As a boy I grew up in the shadow of the Rockies here in southwest Alberta, where the plains come uninterrupted by hills to go swooping up to the spectacular skyline in front of my door. It was a frontier then, a land of boots and saddles, guns and fishing rods, the smell of pines and grass, and the clean, warm feel of the sun in everything. It was wild, free country laced together by crystal streams between the low, folded ridges buttressing the feet of the mountains—all basking under summer skies and glittering coldly in the bitter winds of winter.

The smoke of our camps was a banner against evening skies proclaiming our presence around warm fires redolent of burning willow or pine. Tired from long days on the trail, we sat cross-legged, relaxed and comforted by small things—appreciative of the bare essentials for living—while trading stories of things that happened that day, or maybe a month ago, or even years past. We had no radio or television to steal our time. Most of us saw a newspaper very occasionally and maybe a motion picture once or twice a year. We entertained each other. Sometimes a musician or a singer took his cue under the stars. Mostly we told stories, for storytelling was expected of everyone and those of us who were good at it always enjoyed an enthusiastic and attentive audience. We were to varying degrees elocutionists of a kind; entertainers who drew pictures by colorful word and gesture of the many things we had known or experienced.

A joke on someone was always good for a laugh; the man who told them best being the one who could roar with genuine laughter at one on himself.

So when this book began to take form I found myself wandering with considerable nostalgia down long backtrails —trails that had led me once as a boy and a man through some of the most magnificent country in the world.

And in retrospect I became very much aware that just being born is not enough, for if a man is to enjoy the maximum from living, he must also be endowed with a fair portion of luck in many things: his parents' character and philosophy, his health, environment, and just plain happenstance—the fall and roll of the dice of fate. Most certainly the adventurer born to parents without vision, imagination, humor, and the good sense to allow him to adventure will either leave home too soon with insufficient experience and knowledge to make the most of his life, or he may die wasted, bitter and disillusioned. Real adventurers, the trail blazers and leaders, are made very young, for these attributes are not acquired with the conservative judgment of accumulated years. Like a good horse trained to answer the bit without fighting it so that it can run fast and free with its head held low enough to see the ground and where to put its feet, the youngster who has learned the pitfalls of living early enough without being stifled is one who will enjoy life in full. Surely adventure can take countless forms—there are those who can make selling soap an adventure—but one needs a good measure of health to enjoy it and here again training and luck come into the picture. But perhaps most important of all is growing up where adventure is a part of opening one's eyes after a night of sound sleep, where adventure is so much an integral part of being awake that it shines and beckons every fearless passing moment. For to be fearless, yet respectful of danger, is the golden key.

To tell this story of life on the western frontier I have used many stories told by others, not just to entertain, but to make a point and also to illustrate characters that have had a very definite influence on my life. To these I owe much. It is too late to thank them now. But I can make this book a sort of epitaph that will live on when the sands of time have run out for all of us who trod the wild, free land when it was young. These were people who knew what real values added up to over the long haul; for the most part they were the most generous kind. They were certainly generous to the kid who never missed a chance to wiggle and squirm close to the circle around evening fires when pipes were lit and the stories began to come. They sometimes gave him hell when he did the wrong thing or overstepped his privileges; they were excellent teachers whether they knew it or not; and they helped him gather skills with which to take care of himself in tight places. These were the oldtimers, the explorers, the trail blazers, and empire builders who came into the big country and helped open it without hurting it very much. They lived in it, tough as it was sometimes to stay alive, till their steps faltered and their eyes' keen vision dimmed. Most of them are dead now, a few still share what is left of their living in this country.

To all these I take off my hat in respect and in memory

of those grand days when all the streams were clear and cold and pure as the pristine mountain snow. We trapped the beaver and hunted the wild sheep. We knew the tangy smell of horses and cattle and well-used saddle leather, along with the unforgettable feel of a bucking bronco between one's knees. We gloried in the space and reaches of a great, clean land, where oceans of bunch grass waved in the wind and forests stretched away for miles to timberline backed by a thousand peaks. We shared fires up and down the Rockies and when necessary our blankets.

But as Charlie Russell, the immortal cowboy artist, has said: It doesn't matter much whether a man is headed for heaven or hell; there is always a small set of moccasin tracks ahead of him. For sure no man ever went very far till he was aware of this bit of wise philosophy, for it is woman who makes the life of man possible. While doing it she sometimes smooths and sometimes complicates the way. She puts the love and warmth and light into a home no matter how crude it may be. Through her many sacrifices, she endowed our presence with a permanence; and she brought children and promise to the west.

Although my life has been anything but sheltered, I have known very few really bad women, although some were definitely better than others. Those who have been close to me are a very special kind. All of them being something of a mystery, I do not pretend to know how to write of them in any great volume, so I leave that up to my more learned contemporaries. It is a sure thing that no man is really complete until he has looked deep into the smiling eyes of a good woman while the flames of an evening fire flicker and dance against the roof of his tent or rafters of his house.

I draw some comparisons here in these pages, illustrating the falsity of counting wealth only in bank accounts and not in the natural endowments we enjoy. For the only real wealth is in experience, the appreciation of real beauty, and the friends we gather along the way. It is knowing that a tree is something besides a stack of two-by-fours piled on end waiting to be cut, something more than just an obstruction to a right of way to be bulldozed flat and burned. It is a feeling for a mountain going far beyond the minerals and the water power it may contain. It is knowing that a clean river is something of vastly greater value than a place to dump sewage or a possible location for a dam. It is the deep appreciation of a day in summer when the wild call of an eagle comes down from where it swoops and circles against the white breast of a cloud far above the peaks.

So come with me a while, and I will show you a trail.

# T R A I L S

*of a Wilderness Wanderer*

# By the River

IT IS A PART OF THE TRIBUTARY SYSTEM OF THE MIGHTY Saskatchewan River, which sprawls across the plains of western Canada like a great fallen tree with its branches and twigs caught among the crags and spires of the Rockies. It is one of the smaller rivers that hungrily collect the waters from melting snowfields and glaciers and in turn feed them to the main stream. It has roared and rumbled and murmured sleepily in moods revolving with the seasons for many thousands of years, while countless hordes of living things drank of its life-giving wetness, waded and swam in it, and sometimes drowned in its flood. It has been there since the huge masses of the Ice Age began to retreat and the great plains began to green up after millenniums of frozen silence.

The first drops oozed from the melting ice gathering in pools, and when these pools could hold no more, little rills spilled out to play and join on lower ground to form creeks, which in turn gathered and gamboled noisily with more creeks to mingle in an even bigger stream. So the river was born, predestined by the sun to flow down across the low places in the land, carving the country into hills and valleys to fit its need for running space; and influencing every creeping, walking, running, and flying thing living within its drainage.

Unlike a river, a man is born holding his luck in his two hands, good or bad, instinctively squalling in his helplessness. Regardless of the influence of the sun on his destiny, he has no control in choosing his sire or his dam, and nothing whatever to say about the general geography of his birthplace. His childhood—the beginning rills of his life—can be bank full with joy and love, or maybe drowned in creeks of tears and misery. He may die before he has the chance to taste of living or any part of the independence of choosing his own trail; but regardless of what happens to him, he will owe his life to water, and unless fate decrees that it be melted from a polar ice cap, that water will come from a river. So the lives of men and rivers are tied inexorably and inseparably together, and nothing can change that relationship.

I was born luckier than most—close to this beautiful river in southwest Alberta, first of a second generation born in the same place. The river is called the St. Mary's, and, needless to say, the name is much younger than the stream.

My grandfather came here from the east across the oceans

of grass with an early survey party. He came most of the way by Red River cart before there was any railroad, part of a group of government surveyors preceding the flood of settlers to follow. The year was 1882.

He traveled by boat and rail from Ottawa to Winnipeg, which was then a young city, taking root around the old Hudson's Bay Company trading site of Fort Gary. There he bought a saddle, a .22 caliber single-shot rifle and a .45 caliber Business Sharps, a somewhat lighter version of the Big Fifty Sharps, the favorite rifle of the buffalo hunters. He already had a 12 bore muzzle loading shotgun, so his arsenal was complete, and along with an ax and a few other necessities, he had the basic tools of the frontiersman.

The head of the railroad at that time was Brandon, a few miles across the prairies west of Winnipeg, and it was here that the party outfitted themselves with horses and Red River carts to take them the rest of the way, to the foot of the Rocky Mountains.

The Red River cart was an ingenious and practical contrivance, if not very beautiful, which was invented by the Metis, the French halfbreeds of the Red River Valley. It was a two-wheeled contraption made of tough cottonwood or aspen tied and lashed together with rawhide, sometimes without a nail or a bit of iron in its entire makeup. It was a sort of compromise between an Indian travois and the heavy American freight wagons drawn by bull teams. The wheels were held together and shod with rawhide, which shrinks and grows iron-hard with drying. The rawhide carry-all, slung in a stake basket over the single axle, held the load. It could be drawn by one horse working between shafts or two pulling on each side of a tongue.

No blacksmith or wheelwright was needed when a Red River cart broke down. Grandfather once told me that the only things required to make repairs were a sharp ax, a good knife, a rifle and some native craftsmanship. The ax and the knife were used to carve the replacement piece from the first suitable tree encountered. The rifle was employed to shoot a buffalo, from which the necessary rawhide binding was obtained. When buffalo grew scarce other kinds of hide were used. Most of the time the axles ran dry, for grease collected sand. They ran wood on wood, creaking, howling, and groaning across the prairies with such a din that a string of carts could be heard long before they were sighted. Wet weather was not good for Red River carts, for prolonged soaking made the rawhide go slack and then the vehicles tended to fall apart. They were good for fording streams, however, for they floated like corks and were light enough to be manageable behind swimming horses.

By the time the railroad reached the country just east of the Rockies, my grandfather had taken up a ranch on the bank of the St. Mary's River at the fork by its confluence with Pothole Creek just across from the Blood Indian Reserve. The country was still a wilderness then, without a wire fence between the North Pole and Texas. But changes were in the wind, for no longer did the rifles of the buffalo hunters boom. The buffalo were mostly gone except for a few scattered and bewildered stragglers and a herd of wood bison in the north. It was cow and horse country, where the oldtime cowboy was king.

It was a time of swift transition, for although only a few short years had passed since the first white man had settled here, by the time I was born, fifty-four years ago, the plains were mostly cut up by wire fences. Farmers were turning over the deep prairie sod—plowing under the rich buffalo grass, and the square bulk of grain elevators lifting against the blue of the big sky spoke of the fertility of the soil and sealed the end of the kingdom of the cowboy.

Only the river still swept eastward, unchanged and free. I have an early memory of taking a walk one afternoon with a great aunt when I was still a very small boy. We went out across the flats and climbed up onto the rim of the plains overlooking the valley. We went quietly, without much talk, for my aunt was very deaf, so that it required a shout to make her hear. She was content to say only enough to keep me from falling in a washout or getting into trouble with a prickly-pear cactus. Our relationship was placid, wherein we communicated by looks and touch of hands, and were content with it.

We climbed up a long hogsback ridge to the edge of the prairie above to stand looking back down at the ranch buildings crouched in a scattering of cottonwoods and willows. Under the afternoon sun the river looked like a long silver lariat thrown out in careless curves along the floor of the valley. It came sweeping around a bend at the foot of a high bank a mile above the buildings, slid down a long rapid to a bend at the forks, and then went glinting and dancing along to finally lose itself behind a jutting whaleback ridge away down-valley.

As far as we could see up and down the river, cattle and horses dotted the slopes and flats. Most of them wore my grandfather's brand.

The far rim beyond the valley was flat and on the same level as the spot where we stood. Beyond it, far off, the mountains lifted tall and deep purple in the afternoon sun, somehow mysterious; big and wonderful in a country that was so flat.

Dominant, standing out by itself from the ranks of peaks,

7

was a mountain looking like a great square block. My mother had told me it was called Old Chief, and now I stood gazing at it, wondering as usual why it looked so little like an Indian. My reverie was broken by the sound of Grandfather's big touring car coming along the trail from town. He brought it to a stop in a swirl of yellow dust to pick us up for a ride back to the buildings. For me, riding in that car was a transport to heaven, tremendously exciting, full of strange smells, noises, speed, and great surges of power. I can still see Grandfather sitting with the steering wheel clasped in his powerful hands, his inevitable bent stem pipe gripped in his teeth and his hat sitting on the back of his head. His craggy face glowed with good nature and enjoyment and his eyes gleamed with humor. Being a trained machinist, he was mechanically inclined and owned one of the first cars in the country.

Grandfather fitted into this big country, and when he came he planned to stay. The ranch buildings were standing proof of his intentions. They were constructed of solid concrete and stone with walls and gables a good foot thick. The concrete had been manufactured on the ranch in a kiln he constructed, which had been fired with coal dug close to the site. His method was simple. He dug a chimney-like hole about eighteen feet deep close to the edge of a perpendicular cutbank, which he lined with fieldstone. Digging into this hole from the bottom of the bank, he installed a burning chamber and heavy grates made of railroad iron. The kiln was filled with chunks of almost pure limestone picked up in the immediate area. A supply of coal was dug from a five-foot seam exposed by the river about ten steps from the fire-hole. So he burned lime and manufactured the basic ingredients of concrete from what lay around him on the land. All of the buildings—even the chicken house—were built like fortresses, strong and weatherproof. Some of these buildings still stand in use, proof of the character of the man who built them.

My father built his cottage the same way. It stood about two hundred yards across a coulee from my grandparents' house, and it was here that my brother John and I started out on our respective life trails. Although I did not know it then, mine was destined to follow the river to its head-waters away up among the high crags of the Rockies, and then to wander all over the length and breadth of the wilderness country stretching between the water-carved canyons of Idaho's Salmon River country to vast northern tundra prairies of the Alaska and Yukon Arctic.

Those first steps on the old ranch were a prelude to adventure—not altogether a unique thing, for life is an adventure for all of us. It matters not where we start it, for it is still an adventure. Whether or not we enjoy it depends on some luck, but perhaps more on one's desire to understand and enjoy even the pesky things with which we become inevitably entangled. It is then that the ordinary ceases to be ordinary and the dull takes on a shine, for all things have a story to tell if uncovered and understood.

If someone were to ask me what was the most valuable thing I have inherited from my forebears, I would certainly not list property, but rather an ingrown curiosity about living things and a built-in desire to respect and enjoy the adventure of living. My luck in having an environment in which to fully grasp the fun of being alive made itself manifest early in the game. The everyday happenings at the beginning showed me this before I had walked very far, leaving impressions that still stand paramount.

There was the time I stood on the perimeter of the yard in front of the house, a small figure on the edge of a vast rolling sea of grass, watching in fascination the flickering, ribbon-like gamboling of a weasel. No other animal moves quite like a weasel, for it never walks but always runs, and the running is with a silken poetry of motion—a cadence of muscle and gleaming fur in search of prey.

The weasel was all around me, prying and sniffing into every hole and tuft of grass, paying me not the slightest attention beyond a sharp-eyed stare or two, although it came several times within inches of my bare toes. I did not realize at first that the weasel was hungry and hunting for a meal. All of a sudden it must have run into a mouth-watering scent coming downwind from a chicken coop between me and the house, for it paused with uplifted head and then streaked straight for it.

A fat old hen was living there with a clutch of newly hatched chicks, all placid and self-satisfied until hell arrived and the feathers began to fly. I ran to the coop to kneel and stare spellbound through its slatted front as the weasel began to massacre chicks with all the dispatch of a master killer.

My enchantment with this scene of murder and mayhem was suddenly interrupted by the indignant arrival of my mother carrying a broom. With an angry shriek she upended the coop, turned the survivors loose, and in the same motion took a swipe at the weasel. The little animal dodged the blow with neatness and promptly ran down a gopher hole nearby. The vigor of the blow broke the broom handle, but when the weasel stuck its head up out of the hole to shriek at Mother in return, she attacked it vigorously with what remained of the handle.

Now her weapon was much less unwieldy and her blows quick and accurate, but the weasel moved like flickering light, dodging every swing. The battle was resolved in a stalemate. In no way ready to give up, Mother gave me the club with instructions to keep the murdering little beast in the hole while she went for a bucket of water out of the rain barrel. Her strategy was to drown the weasel out into the open.

It took a lot of water before the hole filled up, leaving the weasel almost totally immersed except for its head and neck, but the little animal did not make the expected break for the open. It just stood eying us sharply and then opened its mouth to squall in defiance. Mother took a swift chop at him with her club. But as usual the weasel ducked like a flash, and the blow spent itself harmlessly. The animal's head came up again with a battle cry.

My father was out riding for stock that morning, so my mother was more or less left to her own devices for eliminating this raider of chicken coops. There is no telling which way the battle would have gone had not the hired man chosen this moment to unexpectedly appear.

It was his day off, and he was dressed in his Sunday best, complete with white shirt, blue serge suit, and polished boots—all ready for his visit to town. Taking in the situation at a glance, he volunteered his services.

"Just hold on a moment, ma'am," he said. "I'll get the boss's gun."

He went into the house and shortly reappeared, fumbling a fat red cartridge into the breech of the shotgun. Whatever his abilities as a hired man, they did not penetrate very far into the dynamics of ballistics and hydraulics; for without undue preamble, he walked right up to the hole. As Mother stepped quickly to one side taking me with her, he deliberately aimed at the weasel at a range of perhaps three feet and blazed away.

It might be truthfully said that he fired into the hole and the next split second the hole fired at him; for what happens when an ounce and a quarter of closely bunched birdshot goes into a gopher tunnel full of water at that range is most impressive, to say the very least. The hired man was transformed into a blinded, dripping mess of mud and water mixed with fragments of weasel. He stood in shock, pawing at his face. A piece of bloody skin to which was attached the bedraggled tail of the weasel hung over one ear. He was a total wreck from head to foot.

My mother and I stood paralyzed, staring at him in wideeyed astonishment. Then, as he began to partially recover, Mother suddenly covered her mouth with a corner of her apron and ran swiftly for the house. Meanwhile he had opened the gun and ejected the spent shell. I saw it lying in the grass and picked it up to curiously examine it. It had an acrid smell of freshly burned powder—a somehow attractive odor that I sniffed with enjoyment as the hired man headed morosely for his quarters.

Ever since that day, when I see some uneducated pilgrim carelessly waving a gun around with no regard for himself or the innocent bystanders, I long to arrange for him to shoot at something in a gopher hole full of water. Nothing could possibly leave anyone with a more lasting impression of the destructive power of a loaded gun.

For some reason I sometimes still find myself sniffing absentmindedly at the open end of a fired shell with unexplainable enjoyment along with a vivid memory of the weasel that took shelter in a hole.

Early life along the bank of the river was full of firsttime things one remembers best. Someone once said that the three greatest adventures of life are being born, becoming married, and dying. Nobody has as yet commented on the comparative quality of the last from first-hand experience;

but there are a lot of unlisted chapters in between, worthy of mention. Certainly time and geography were kind to my brother and me; for just being born on the edge of the frontier when it was changing so rapidly was in itself something to see, and the seeing was greatly enhanced by being part of a family that had so much a part in it. Like all people, we had our share of disappointments, frustrations, hardships, and tragedy; but it was not a family characteristic to spend undue time dwelling in self-pity or indulging in the luxury of prolonged attempts to lay the blame for our misfortunes on others.

The river was an influence in its interminable journey downstream—always ready to provide us with water when we needed it, a boundary marker for the ranch on its western side, always moving, except in winter—a sound reminding us day and night of the inevitable journey of life. In summer the sound was like a lifting and falling lyric, sweeping down from between the hills. In winter the stream snapped and boomed in its shackles of ice, contracting and expanding in the changing temperatures. At break-up time in spring it put on a show to leave one shaken with the power of it; for as the warming sun rotted the ice and melting snow swelled the side streams running into it, there was a building up of pressure to a point of sudden bursting. At some hour of the day or night there would be a sudden cannonading of breaking ice that echoed off the banks, and like magic the whole surface of the river would be on the move with legions of ice blocks of every size, shape, and form; heaving, grinding, and gouging as they swept away downstream. Sometimes the blocks would jam, forming temporary dams that would back the water up, flooding bars and flats. Then the pressure would smash its way through and just as rapidly the waters would recede, leaving acres of stranded ice blocks high and dry—incongruous-looking baby icebergs lost from their element, crying sadly in the sun.

The river was a constant provider for men and stock, but it took its toll among the unwary, the young, and the inexperienced. Almost every year somewhere along its length there came a story of tragedy. The survivors were strong and fearless swimmers with a built-in brand of courage no longer so evident among prairie dwellers.

My father tells of an experience encountered at an age when most modern youth have never had to make a vital decision and take their exercise riding a machine.

In 1907 he and a cousin, Joe Bell, were coming back from a spring round-up in the Little Bow country. They were driving a gather of cattle wearing my grandfather's brand, as well as their *remuda* of eighteen or twenty saddle horses, one of which was packed with their warbags and bedrolls, when they ran into a flood. The rivers were out over their banks where they came to cross at the forks of the Waterton and the Belly, swimming deep from bank to bank with some flooded cottonwood timber in between. Waiting for the water to go down was out of the question, for the men's grub was running low and they were anxious to get home. So they took off their chaps and boots and tied them behind their saddles and made ready for a big swim.

First they drove the loose horses into the water and then the cattle. As they rode in behind the tail of their herd their saddle horses were at swimming depth in two steps. Both riders were ready, their cinches loosened as always before hitting deep water, for a tightly cinched saddle horse will sink. If their horses got into trouble, they could slip off and take the mount by the tail, for a horse can swim better this way and he can still be steered. Furthermore, when shore was reached the rider would be with his horse. Most drownings occurred when a man panicked and quit his horse or was not properly prepared for swimming water. But there was no trouble in this instance, and in no time they were across the raging flood. But they still had one more river to cross before they could sleep under a roof.

It was midnight when they drove their stock down off the hills to the edge of the St. Mary's, and it was in full flood. They had a choice of sleeping out in a steady drizzle in bedrolls soaked from the first crossing or heading into the river. Sopping bedrolls held small promise of comfort and it was too wet to build a fire and cook some supper; so again they drove the horses and cattle into the river and headed across. My father comments: "There is no more lonesome place in the world than swimming a flooded river at midnight, but we made it. We had good horses that had been swimming rivers since they were suckers." They had been swimming rivers since not long past being suckers themselves, which no doubt contributed to the fact of two successful crossings in one day; for indecision or fear could be fatal in such places, especially swimming in the dark of midnight.

When I was four years old and my brother John was still a babe in arms, my father moved us to a new ranch away up on a tributary called Drywood Creek at the foot of the Rockies. Here we came to know the mountains, and here the stream was small enough to wade most of the time, as well as being so icy cold that it gave us small chance to learn to swim. But we came back in summer to visit our grandparents and there we learned to swim in the river.

I remember standing on a ledge one day, just off the edge of deep water, fishing for whatever would take a minnow on a hook. Suddenly there was an explosion at the end of my line—a smashing jerk that almost tore the willow pole from my hands. Before I knew what was happening, I was thrown off balance and yanked bodily into the river over my head. I came up spluttering and spitting out river water, more surprised than scared, with the current sweeping me downstream along the foot of an almost perpendicular rocky bank. There was no place to grab hold and climb out for some distance and no choice but to swim. Up to that point all the swimming I had done was dog-paddling across quiet places where I could touch bottom with my toes. This was for keeps. I kicked and aimed for a point jutting out into the river downstream. It seemed like no time till I had a solid hold and was pulling myself out on the bank mighty glad to be free of the river. My rod had disappeared and I never did find out what kind of monster had grabbed my hook.

When I came into the house a bit later wet to the top of my head, dripping water all over the floor, my grandmother eyed me sharply and asked what had happened. If it had been my mother I might have lied a bit to save her some worry and fuss, but it was no use lying to my grandmother. She had a second sight—a kind of magic vision that allowed her to see right through people—especially boys with whom she had accumulated an uncommon amount of experience.

So I told her. "A big fish grabbed my hook and pulled me in the river."

"Hmm," said my grandmother. "How did you get out?"

"I swam out," I told her proudly.

She looked at me keenly and silently for a moment before saying, "You must practice swimming until you can swim anywhere. It would be very difficult for me to explain to your mother if one of you drowned."

There were no warnings—no fuss. Just plain commonsense advice from a woman who knew what it meant to survive on the frontier.

# BY THE RIVER

I was ten that summer and my brother was six. Being four years younger he learned to do everything just about four years sooner than I. Following Grandmother's advice we practiced swimming till we were worked down to rope-thin, sun-browned little savages. Then one day we tackled the river all alone.

I will never forget that swim.

It was a blazing hot day when we came to the edge of a big sand bar between a strip of willows and the water. We stood looking at the water slipping past, cool and inviting in lazy swirls. Across the river at the foot of a three-hundred-foot-high bank deep, cool shade beckoned where a sandy cove cut a bay into the bank. Without any talk we slipped out of our clothes to plunge into the current and begin to swim across.

At this point the river was about seventy-five yards wide with a fairly easy current. We were halfway across before we began to feel the effort, but it didn't bother us at first. My brother was swimming a bit upstream and just ahead of me, with his feet stirring the water about even with my shoulder.

Another few yards brought the far bank closer but then we hit faster current and the distance began to tell. My arms felt heavy and hard to pull through the water. For a long moment I wondered if we were going to make it and for a second I faltered in what could have built into big trouble. The difference between being obliterated and surviving the first real test of danger is the ability to recognize it, dodge, and keep going. John was still ahead of me, his stroke still steady but slower. The current had swung him past me so that he was drifting downstream with it. I turned to follow, picking up my stroke. Almost magically the going was much easier and then I felt sand under my knees. We were across.

There was an unforgettable feeling of relief and triumph all mixed up. We didn't have much to say to each other for want of the wind to say it. We knew without saying that we had been flirting with disaster. We sat naked on the sand, shivering a bit, hugging our knees and soberly contemplating the way back. It was one thing to swim the river, but now we were on the far side without our clothes and no place to go. It never occurred to us to just stay there and wait till someone came looking for us. That would be an unforgivable concession to fear. We didn't even contemplate the possibility that we might be in a very real predicament. We just sat and looked at the river going by, wondering how it was going to be on the way back.

But the river had taught us something valuable, for when we were rested and ready for the return, we walked upstream a ways to a little gravel bar jutting out into the current, and from its point we began to swim side by side on a long slant downstream. Coming back was so easy we hardly took a deep breath. As we put on our clothes, enjoying the heat of the sun, we both knew we could swim. What was more important, perhaps, was that we had learned to go with the river and not fight it.

We grew as we learned; hearing, looking, sniffing, and tasting things there on my grandfather's ranch. There was the bawling of cattle along with the smell of dust and burning hair at branding time. There was the breath-catching sight of lifting banners of dust against the sky as the horse herd came thundering down into the valley, being driven toward the corrals. There was the stirring picture of a cowboy topping off a bronc, sitting tall and lithe in the saddle with spurs flashing forward and back in time to the bone-jarring twisting and bucking. Quite often we saw an uncle or my father making a lariat do things that only a master can call on it to perform. We chased jackrabbits and coyotes with hounds. There was hunting and fishing to satisfy anyone, and the larder rarely lacked a variety of things that the country could provide.

We came to know something of animals. We came to know something of people by meeting the procession of visitors passing through. We listened to their stories and thus came to find out some of the things motivating them in their quest. Becoming familiar with all kinds of warm-blooded things like ourselves, living with them and learning from them, opened a great door that has never closed.

2

# *The Ashes of Old Fires*

THE PEOPLE WHO FIRST LIVED IN THE VALLEYS AND PLAINS of the Saskatchewan River drainage appear to have come from the north thousands of years ago during the Ice Ages. There were four of these glacial periods. During the first three, what is now Alberta was a part of a great open corridor—what is known as a refugium—an ice-free region that apparently stretched north and west throughout the Northwest territories, Yukon, and Alaska. It abounded with plant growth and consequently with wild game, which in turn attracted man across the land bridge leading from eastern Siberia into Alaska via the Seward Peninsula and the mountains now mapped as the Aleutian Islands. Because so much of the latent ocean water was then inland, in the form of ice, this land bridge was then high and dry.

The last of the Ice Ages, of which we can still find remnants scattered through the Rockies and coastal ranges to the west and north, was much more widespread, covering all of Alberta, except for a few ranges of hills, to a depth of perhaps five thousand feet. One ice mass originated from a great dominating dome near Hudson's Bay. It met another spreading out from north central British Columbia, where only the tops of the highest peaks stood clear of the ice. Together these two mammoth ice masses ground their way slowly southward, pushing all living things ahead of them. When a warming climatic trend finally sent this glacial time into retreat, men and animals followed it back to the north, and thus the region of the Saskatchewan was repopulated.

These old trails of men following their noses into new country and rich hunting grounds are well tracked by many things left behind—spear points, arrowheads, the charcoal and ashes of ancient campfires and fragments of the bones of animals they hunted. The origin of these peoples is marked by their Mongolian features still plainly evident among many of the western tribes today. Even the manner of speech of some of the tribes puts blazes on an ancient migration road running from Alaska to Mexico—a route described by historian Dan Cushman as The Great North Trail.

For except for slight differences in dialect, the Hareskin Indians, a sub-tribe of the Slaveys, living just north of the Arctic Circle and east of the Mackenzie River, speak almost exactly the same language as the Navajos of the New Mexico and Arizona desert country far to the south. A few years ago while on a northern exploration expedition I visited a Hareskin village on the shore of Coleville Lake. There I met and talked with Father Bernard Brown, who had just finished building a beautiful Our Lady of the Snows Mission there—a truly remarkable structure of logs in a land where suitable trees are scarce. He is one of the very few white men who has mastered the Hareskin tongue and has made a considerable study of these interesting and attractive people. Upon hearing from a visitor that the Hareskin language closely resembled that of the Navajos, he made a long trip south to visit that tribe and completely astonished them by speaking to them in Hareskin, which they understood.

Apparently the two tribes were originally one, but during the course of an ancient migration they split apart for some reason. One part had stayed on the frozen prairies of the Mackenzie River Basin, while the other drifted far south into the arid desert regions.

There were other splits of a more recent nature among the tribes. The Sarcees of the south Alberta foothill country were originally a group that broke away from the powerful Beaver tribe of the upper Peace River and Liard River drainages of northern British Columbia. They moved south into the Blackfoot hunting grounds and when that tribe found these intruders too tough and persistent to be driven out, they made a treaty with them, whereby the Sarcees became a part of the Blackfoot nation about 1790. The Stonies, Rocky Mountain hunters who ranged between the Kootenai and Blackfoot territories, were an offshoot from the Assiniboins of Manitoba. They apparently came west after they acquired horses and established their hunting grounds in the rugged mountains along the Continental Divide.

Prior to the arrival of the Spaniards in North America, these upper Saskatchewan tribesmen were afoot like all the rest of the Indians on the great plains. They lived in a vigorous climate of a vast wilderness teeming with buffalo, elk, antelope, moose, deer, bighorn sheep, mountain goats, grizzly and black bears, as well as many species of smaller wildlife.

The only domestic animals they employed were their dogs, which were used to pull small travois and carry packs, thus assisting the women when camp was moved from one location to another. Life is never dull when dogs are used in this fashion, for the use of leashes was not practical and the canines ran free. From personal experience with pack dogs in the Yukon, I can attest that they leave much to be desired as pack animals, although they can travel over incredibly rough country and carry a load equal to half their own weight—proportionately far more than a horse.

One can picture a band of Indians moving across the plains in those days, the young men striding proudly in front carrying nothing but their weapons, the women strung out behind them, loaded down with tribal goods and possessions, and the whole lot accompanied by a motley mixture of children and dogs—the latter burdened with travois and packs.

Inevitably, game of one kind or another was sighted, and the women were hard-pressed to keep order among the dogs. A jackrabbit approached by such a procession quite often freezes in his hide until almost stepped on, whereupon he goes streaking off with rump bounding provocatively and ears standing erect. Any normal dog is incapable of resisting such a temptation, especially when accompanied by children, for there is a sympathetic bond between dogs and young humans that can be volcanic. In such moments there was a bedlam of shrieking women and yelping canines and children. The earth fairly trembled with vibrations. Under such circumstances an enterprising dog can find six or seven ways to shed its pack in about half as many seconds, and even if he fails to run out from under his load in twenty jumps, there are uproar and confusion to burn. Moving camp in the old "dog days" was well spiced with action, harried women, and much searching for scattered belongings. No doubt many a dry camp was made as a result of time lost, for Indians have always been loath to travel after dark. As a consequence of hazards and the hard work of travel, Indian camps in those days were of a semi-permanent nature.

Theirs was a true communal life, wherein there were no very rich people or any very poor. They killed hundreds of buffalo at a time by driving them off steep bluffs and cliffs; or making a surround in a corral or some natural feature of ground, thereby maneuvering a great milling herd to a relative standstill while the kill was made. Everyone shared alike, and hunger was practically unknown except during occasional severe weather in winter.

But then about 1700 came the horses—descendants of those lost by early Spanish explorers and settlers—and within a very few years the way of life of the plains Indians changed remarkably. In a short time they became almost completely nomadic, drifting from place to place with their colorful skin lodges, hunting and fighting on horseback, and in general developing into a proud, arrogant people that would be listed by historians of the early west as among the finest cavalry of the world.

Wealth came to be measured in horses, and because of the inevitable differences in individual characteristics among people, the rich became richer and the poor even poorer. With time on their hands, the men, like horsemen the world over, developed a distinctive dress, which was spectacular in buckskin, eagle feathers, and brightly dyed porcupine-quill embroidery trimmed sometimes with shells. This buffalo hunter cut a picturesque figure sitting proudly on his favorite horse, all decked out with his lance, bow and arrows, medicine bag, and other finery—all trimmed with fringes and eagle feathers and paint. Such early western artists as Remington and Russell captured his spirit well.

Never from the time of my first memories do I recall any member of our family looking down with scorn at the Indians. There were times when we were bound to feel pity for them, degraded as they have been by a progression of errors fostered in the first place by a well-meaning government and maintained ever since by a top-heavy bureaucracy. We recognized them as people with ways that were different from ours. Grandfather's ranch was located across the St. Mary's River from the Blood Reserve, so meeting Indians was something my brother and I grew up with and recognized as part of our world.

By that time the wagon had replaced the travois, although the distinctive ruts left by the dragging ends of the poles were still clearly visible on the prairie. It had been only a few years since the Indians had been confined to reserves, but the Bloods were no longer hunters. They wore their traditional skin clothes only on ceremonial occasions. The women draped themselves in blankets, while the men wore nondescript white man's clothing. Most of them still used moccasins; but the feathered and furred headdresses had been largely replaced by hats. The men still wore their hair in long braids.

Heading for town in a wagon pulled by a pair of undersized cayuses, an Indian family group was a forlorn comparison to the oldtime prairie people. But when they were mounted on their ponies, there was a bit of the old sparkle of pride and arrogance. For horses still represented wealth among them, and by this standard some Indians were rich.

The Indian was all man when he sat tall in the saddle. A few excelled as horsemen, competing successfully with the best of the white men in contests of the open range. These were the remnants—the throwbacks to the much-feared wild cavalry met on the plains by the first explorers and settlers.

The Bloods were a powerful part of those Indian tribal units encountered. The first successful trading with them was carried out by the Hudson's Bay Company from posts scattered thinly on the edges of their range. Like all Indians the Blackfeet succumbed to the lure of alcoholic drink. The Hudson's Bay traders were after beaver skins, but these proud warriors considered trapping "squaw work," and obtained the goods for their trade by raiding American trappers and traders to the south along the upper reaches of the Missouri. If the Hudson's Bay Company was aware of the source of the furs the Blackfeet brought to trade, they chose to ignore it, but there was little love lost between the Indians and the traders.

The Blackfeet, Bloods, Peigans, and Sarcees—all members of the Blackfoot nation—were deeply religious people in their own way, and they had many elaborate rituals. One of these was the arduous Sun Dance ceremony, during which the youths proved themselves men fit to take their places among warriors.

A pole was set in the ground, surrounded by a ring of stakes a bit taller than a man with poles tied between them and the high center pole. Loose boughs and brush were thrown on this latticework of poles to make a sort of crude dancing lodge, where the people sat and stood in a ring around the outer perimeter.

The ceremony opened when the youths came one by one into the watching ring of people. Old women took sharp knives and cut slits in each man's breast muscles. The medicine men lifted the strips with sticks and passed rawhide thongs beneath the muscles, knotting them firmly, and tied the other end of the thongs to the top of the center pole. Then the young Indians danced and repeatedly threw themselves back with all their weight against the thongs until they tore free of the skin and muscle holding them.

Another method was to slit the muscles of the back over each shoulder blade, tying short rawhide thongs to these flesh loops. Heavy buffalo skulls were tied to the thongs to drag behind the young dancer as he tried to break free. Sometimes a friend or a relation would hold back on the skull, giving him a better chance to tear out the restraining loops of flesh on his back.

When I was a boy I saw Indians from the Blood Reserve who still carried the scars of the "making brave" ceremony of the Sun Dance, but they are all gone now.


NOT MUCH IS REALLY KNOWN of the country lying between the North and South Saskatchewan Rivers previous to the beginning of the last century, for it was the last territory penetrated by white men in the great plains region. Those that did explore it first did not leave much by way of descriptive records. It is thought that about one hundred thousand Indians lived between what is now the International Border and the Arctic Circle from the mountains east to a line bisecting the forks of the river. It was the last great stronghold of the Indians, until late in the nineteenth century when the Blackfoot brought smallpox from the Missouri River country. This frightful scourge went through the tribes like a prairie fire, for they had no resistance to any of the white man's infectious diseases, and consequently they died in great numbers, giving the traders and settlers the opportunity to take and hold control.

Although there were many times when the tempers of the Indians were a short fuse requiring very little to touch them off into an explosion, there was comparatively little bloodshed between white settlers and the red men during the time of early settlement in western Canada. Unlike the western States, where the settlers carried the law in a belt holster, and Indian treaties were often broken before the ink was thoroughly dry, the law came early into western Canada, enforced by men in the uniform of the Royal Canadian Northwest Mounted Police. White man and Indian were, for the most part, treated impartially by this small force of intrepid officers, who patrolled a vast region and who by their very lack of numbers and exercise of sheer nerve impressed the Indians and largely won their respect.

In 1874 two troops of the newly formed Royal Canadian Northwest Mounted Police, under the command of Colonel J. F. Macleod, rode across the plains heading for the mountains to the west. Their first item of duty was to take Fort Whoop-up, a whisky trading headquarters on the forks of the St. Mary's and Belly Rivers about six miles below the future site of my grandfather's ranch. The fort had been strongly constructed out of hand-hewn cottonwood logs by American traders under the command of a man named Hamilton in the late 1860's. It was the kingpin of the illicit whisky trade in the region. A desperate battle was anticipated by the police, for the fort was known to be heavily manned and armed with a cannon.

As they advanced across the vast rolling emptiness of the plains, this small force of 274 officers and enlisted men was never quite sure when a thousand mounted Indians might suddenly come boiling out of a fold of ground to overwhelm them. This, on top of the anticipated battle at the fort, made them understandably nervous. One day after many miles had been covered and it was evident they were getting close to the fort, one of the officers asked Jerry Potts, their half-Indian guide, what they would see upon topping the next hill. That worthy of the plains was not very much comfort. He looked sharply at his questioner, gazed reflectively at the country ahead, spat in the grass, and said, "Nudder hill!"

Finally they arrived on the high ground above the in-famous Fort Whoop-up, primed for trouble. They recon-noitered the place with great care, placed their pair of field guns in advantageous positions covering the place, and then a party was dispatched to ride down to the gates and demand surrender. The loud knocking reverberated hollowly through the fort, followed by long pregnant moments of silence. Then there came a slight shuffling, followed by noises of the inner bar being lifted, and an ancient half-Indian caretaker opened the gate a crack to peek out with startled eyes at what must have looked to him like hordes of redcoats armed to the teeth. He broke cover and ran like a startled buck to vanish into the willows along the river, and was never seen or heard from again by the police. Fort Whoop-up was thus captured without a single shot being fired.

Colonel Macleod led his men west and proceeded to build a fort on an island in the middle of the Oldman River not far from where the town of Fort Macleod now stands. From this headquarters stemmed the detachments and patrols that established law and order in the Canadian west.

As a family we were more than just bystanders watching a time of swift and overwhelming change of Indian life on the prairies; we were a keenly interested part of it. We were always aware that what we called law and order was too often the exact opposite by Indian standards. There were innumerable times when the Indians saw the white men break their own laws and be generally just about as hypocritical as people can get. Too often white men got away with their lawlessness, which made the Indians wonder and smolder—for in the not too distant past the Indian who broke tribal law was quickly and severely punished by the Dog Soldiers, the Blackfoot tribal police force of the prairie.

The Indians appreciated a firm, fair hand in law enforcement, and it was not always necessary for the Mounted Police to use force. Once, when a chief rode across the river from an Indian encampment to dismount and brazenly lift a colorful braided rag rug being aired out on the top rail of the yard fence, my diminutive grandmother ran out of the house with a broom and caught up to him before he had a chance to get back on his horse. She did not take the rug from him, but with flashing eyes ordered him to put it back. Faced with a little woman, then twenty, who weighed only a hundred pounds, he was naturally reluctant to obey, for the whole camp was watching. When she swatted him across the rump with the broom and repeated the order, he was burning with anger but mighty surprised too, for who was this little white woman to strike a chief? He was faced with a dilemma, for he could not very well strike back without losing face and also inviting trouble from my grandfather, nor could he obey without more of the same. Grandma never gave him a chance to get organized. She drove him to the fence, where he threw the rug back where it belonged and quickly retreated, to the huge amusement of the watching Indians, who thought it was a great joke on him.

Although she would never feed an Indian in her house when my grandfather was away from home, Grandmother was always kind to the Indians. By being fair and completely fearless, she won their respect and held it till the day she died.

This process of carving a home and making a living in the midst of a vast prairie wilderness was far from being a tame business, for the Indians, though confined to reserves, were still wild; nor could the white men who came to make use of the grass be always classed as gentle, mild-mannered individuals. They were by the nature of their environment a tough, active, adventurous, and sometimes unscrupulous breed with a salty sense of humor on occasion.

When my grandfather first came to Fort Macleod in 1882 while working on the survey, he made the acquaintance of the local blacksmith, a man known as Smiley. The biggest part of Smiley's business was shoeing police mounts and repairing the ironwork and chains that went with the big freight wagons and bull teams owned by the I. J. Baker Company. This company hauled all the freight from the steamboats on the Missouri at Fort Benton in Montana, which was the main source of goods coming into the country at that time. Smiley prevailed on my grandfather for a loan, which was given to him upon transfer of a note. When the loan came due, Grandfather went to see him one snappy cold morning in November. He found the blacksmith asleep in his shop, snoring and dead to the world on a work bench. It turned out Smiley had a drinking problem. Heading for his shack the previous evening very drunk, he apparently became a little confused and went to his shop. There, in lieu of going to bed, he lay down on the bench beside a pile of chains. It was near zero, and as the cold penetrated, he began hauling chains across his middle. When Grandfather found him he was snoring under enough chains to load down a big mule and blue with cold.

To solve the problem of the loan, Grandfather took over the shop and hired Smiley to work for him. He held back part of his wages until the loan was paid off, and then gave the shop back to him.

Cattlemen were divided into two classes. One kind owned stock legitimately acquired. The other kind were long-rope artists who stole cattle and horses belonging to other people. The latter were sometimes apprehended and sent away for some educational adjustment in jail; but occasionally luck and cunning delayed the process longer than the community at large would like.

One of these men went by the name of Hippo Johnson. He got his first name not because he in any way resembled a hippopotamus but because he branded his cattle with a big O burned around a hip joint. So his brand was called Hip O, which quite naturally led to the nickname.

There was a man by the name of Maunsell who had a big ranch and a butcher shop at Fort Macleod. Maunsell was an Englishman of fairly recent extraction, and while he was a reasonably successful businessman, he did not have what might be termed by his neighbors as "cow savvy." To make up for this shortcoming he had an excellent foreman—the kind who knew what a cow said to her calf. This man had reason to be highly suspicious of Hippo Johnson's activities in the dark of the moon, and Johnson knew it.

One day when Johnson was riding into town he came upon a big four-year-old calico-colored steer in a draw, and he immediately saw an opportunity for a joke on his enemy. So he drove the animal into a small corral back of Maunsell's butcher shop. Dismounting, he walked into the shop, where the proprietor was busy cutting meat for a customer, and suggested innocently that he should come have a look at a steer out back in the corral. This Maunsell did, in due course, but failed to notice that the steer was wearing his brand. This may have been at least partly due to the fact that Johnson quickly and carefully maneuvered the steer so the brand was hidden against the fence.

The ensuing conversation was salutary and went something like this:

"Good-looking steer," remarked Maunsell.

"Yeah, I thought you might like him," from the dead-pan Johnson.

"I could use a steer like that right now. How much do you want for him?"

"How much will you give?"

"Going price. Forty dollars."

"Good enough. No rush about the money. I'll come back later for it."

"Here, take the money now. I'll make out a bill of sale. If you're in a hurry, you can come back for it."

So with forty dollars in pocket and full of suppressed laughter, the incomparable Johnson made his way to the nearest bar.

Later in the day Maunsell's foreman rode in from the ranch. When he took his horse into the corral back of the butcher shop, he saw the fresh, familiar-looking hide slung on the fence. He examined it and confirmed the fact that it wore the Maunsell brand. Somewhat mystified, for he usually picked and drove in the steers to be slaughtered, he walked into the shop to speak to his boss.

"I see you've killed a steer," he remarked.

"Yes," affirmed Maunsell. "I wasn't sure when you would get in from the ranch, and I needed the meat, so I bought a steer from Johnson."

"You what! Good God, man, that steer was wearing your brand!"

Before Maunsell could find anything to say further, the foreman went storming out to look for Johnson. When he found that worthy in the bar, he barely managed to restrain an impulse to kill him on the spot, but if Johnson was aware of his danger, he never batted an eye. When he found himself confronted with the accusation of being a dirty, unprincipled, no-good, so-and-so of a cow thief, he tipped his hat back, grinned at his accuser and remarked that it was not his fault if Maunsell wanted to buy one of his own steers. Going into great detail, he unfolded the caper from beginning to end, especially reiterating the fact that he, Johnson, had never claimed the steer, and that he had just offered to show it to Maunsell. "If your boss wanted to buy his own steer, I sure wasn't going to stand in his way," he guffawed. Needless to say, the surrounding audience in the saloon was delighted with the joke, and when Johnson insisted that the foreman take back the forty dollars, "just so there wouldn't be any misunderstanding or hard feelings," they were even more appreciative.

It made a colorful story that has gone the rounds among oldtimers for over seventy years. I heard it from an old cowboy forty years ago. This was frontier humor and entertainment, before people started writing implausible themes for western television shows. To be sure, somebody was usually the "goat," and the measure of a man was at least partly gauged by his ability to laugh at a joke on himself.

There was the time Cap Thomas and a friend joined in a poker game with a couple of strangers at an all-night session at Pincher Station—the station on the railroad about two miles north of the town of Pincher Creek. Because they were two miles from the police detachment the boys were wearing their sixshooters, weapons usually confined to the open range on the Canadian side of the Border, for the Mounted Police forbade wearing guns in town.

The game was going steadily against Cap and his partner. Cap was suspicious that the cards were being stacked, but he gave no hint of his distrust and kept on stubbornly and steadily losing money. Finally, in the small hours of the morning, one of the strangers made a slight slip in his dealing, whereupon the hot-tempered Cap pounced on him with an accusation of cheating. The man reared back in his chair as though reaching for his gun, which was a great mistake, for Cap promptly pulled his big sixshooter and shot the erring card player out of his chair.

The man never moved from where he hit the floor, and blood was running on the boards as Cap backed out the door with his smoking pistol in his hand. Sure that he had killed the man, shocked and scared, and knowing the police would soon be on his trail, Cap stepped up on his horse and burned the wind for his little ranch.

Cap was a young bachelor, slim and handsome—somewhat of a dandy, fond of well-tailored clothes. His boots were handmade. His saddle and horse jewelry ran to fine flower tooling and sterling-silver inlays.

When he reached home shortly before sunup, he paused only long enough to throw some grub in a sack, roll up a couple of blankets in a light tarp, and catch his top horse out of the pasture. Then he rode west and north to lose himself in the timbered breaks of the Porcupine Hills.

And so began a period in Cap's life when he learned something of the price a man can be forced to pay for freedom. Instead of a carefree and somewhat reckless young cowboy, who took his fun where he found it, he was now one of the hunted. At night he brooded and shivered beside a tiny fire built in some remote hole in the hills, where it was unlikely to be seen by anyone beside himself. By day he lived on "the dodge," letting no man come close and being continually on the alert for police patrols.

He grew gaunt and hollow-eyed with a dark stubble of whiskers covering his usually clean-shaven face. His fine clothes became dirty and ragged. Only his ivory-handled Colt was kept clean and well oiled. By contrast his horse stayed sleek and fat, for it was early summer and the feed was plentiful and lush.

Cap teetered precariously between giving himself up and going completely bad as he worked his way south to a remote timbered basin near the base of Old Chief Mountain not far from the International Border. He was sitting his horse among some scrub pines on top of a butte one bright morning when he spotted a rider coming up along a little creek across a chain of open meadows. It was the familiar figure of his friend who had been with him on that fateful night of the shooting at Pincher Station. Unable to resist the opportunity for some talk, and also hoping for some food, Cap rode down to intercept him.

When they met, they just sat their horses for a while saying nothing, and then Cap's friend began to swear softly, fluently, and with great feeling. In language well spiced with adjectives reflecting on Cap's intelligence, his principles, and general ancestry for generations back, Cap was told how for weeks his friend had been searching for him to no avail, and how tired he was of looking after Cap's stock and ranch between sessions of looking. Finally he got around to telling Cap that all his running and hiding had been for naught, for he had not killed the man after all. The bullet had just creased him over an ear and knocked him unconscious. Upon some application of cold water, the man had recovered his senses and disappeared without reporting his wound to the police.

When Cap told me this story sixty years later, he remarked, "The joke was sure on me! But every time I tried to laugh about it for a long time after, my face ached." He opinioned, as a sort of conclusion, that a man was a damn fool to carry a gun unless somebody was threatening to shoot him.

BECAUSE THE BLACKFEET were not agreeable to white men penetrating their range, and were a sufficiently numerous and warlike force to make their stand and keep it, theirs was the last piece of country lost to the settlers. Even the early free trappers, about as tough and uncompromising a breed as was ever developed on any frontier, did not penetrate Blackfoot country with much hope of staying. Although they were a product of the survival of the fittest, masters at reading sign, and tough as rawhide—mentally and physically the match for any Indian—the Blackfeet were their nemesis. The bones of many trappers who dared poach on Blackfoot hunting grounds were scattered by animals, and bird feathers and trappers' hair went to decorate the battle dress of the warriors.

But some survived, and one of these I met when I was a small boy; he left an unforgettable impression on my mind. He was then somewhere in his sixties, the kind of man that age does not touch with a very heavy hand. His hair was raven-black except for a sprinkling of silver at the temples and his step was lithe and deceptively quick. Although I did not fully appreciate it then, he was the last of the mountain free trappers, a holdover from old times who knew woodcraft and mountains to the point of it being a part of him.

He was half-Indian—likely Algonquin from Ontario—and as a young man about twenty years old, he came west into Blackfoot country and set up his trapping headquarters on the headwaters of Belly River in Montana. His name was Joe Cosley. He was a crack shot with both rifle and pistol, a first-class canoe man, and an excellent horseman. But what made him most unusual was the fact that he was college-educated. Even by white man's standards of those times he was a highly educated man. He spoke English fluently in a quiet, authoritative manner with a fine choice of words. He was tall and striking in appearance. Perhaps because he chose his trapping grounds in a valley system little used by the Blackfeet—a piece of territory on the American side of the forty-ninth parallel partly drained by waters running into the Saskatchewan to the north—he was tolerated and allowed to stay.

The country he trapped was fur rich. It was cradled among some of the most rugged and magnificent peaks of the Rockies, where snow lay from six to twelve feet deep in winter. Sufficient unto himself, tough and completely the master of his environment, Joe rode his long snowshoes through it winter after winter. He continued to harvest fur there until this section of northern Montana was set aside by an act of Congress to become Glacier National Park. Then the American officials sent to administer the new park informed Joe that he would be required to look elsewhere for a trapping country.

Joe shrugged and moved out, but he did not go far. He simply moved over to the Canadian side of the line, where he proceeded to hunt and trap in the Flathead region in south-west British Columbia—an area adjacent to the northwest section of Glacier Park. In the spring he often slipped back into one of the high valleys on the American side of the line, where he proceeded to lift a fine catch of beaver—a turn of events that the rangers usually failed to discover until he was long gone back to his Canadian headquarters.

One spring he went back to his old haunts up on the head of Belly River, where he made a rich haul of beaver. About break-up time, he cached the furs in a secret hide, and made his way across the mountains to Kalispel, a small town on the Flathead River west of Marias Pass in Montana, where he planned to contact a fur buyer and make arrangements to sell his catch.

Unknown to him, someone who was aware that he had been poaching in Glacier Park had reported him to the authorities there. So when he appeared in Kalispel, a watch was set to observe his activities. The plan was to let him make arrangements to sell his fur and then trail him back to his fur cache, where an arrest would be made and Joe would be forced to pay for his lighthearted treatment of Park regulations. But Joe was not the kind who went to sleep on his feet anywhere. He was as keen as a knife blade and alert to everything that went on around him. Furthermore he had many friends, and perhaps one of these tipped him off to his danger.

Sizing up the situation quickly, he changed his plans and slipped out of town one early morning, heading back over the mountains via Bowman Pass toward the head of Waterton Lakes. The Park Rangers spotted him going, and knowing they had no chance of catching him in the mountains, they quickly took to the rough trail going over Marias Pass in a brand-new innovation in Park equipment, an automobile. Their plan was to circle to the east of the mountains up into Canada via Cardston, Alberta, circle back into the

Belly River country and intercept Joe when he emptied his cache. It was a long way by the road. There were many mudholes to negotiate. The rangers in the car had their work cut out for them.

In the meantime Joe was on the move. Maybe he had spotted the car heading out toward the pass from Kalispel. Perhaps he had been aware of it all along and guessed what was going to happen. At any rate, he set out to race the Park Rangers to Belly River. All day he walked up and over the snow-choked, avalanche-ridden reaches of Bowman Pass and down the east side into the headwaters of Kootenai River. Stopping only briefly to boil the kettle and take on some food, he raced nonstop for about fifty of the toughest kind of miles, arriving in Waterton a bit before midnight. Tapping on the windowpane of a cabin belonging to a friend, he quickly borrowed a small canoe, placed it in the water at Emerald Bay at the foot of the upper lake, and headed east. Caching the canoe in the brush on the east end of Knight's Lake, he headed out again on a long trot. At dawn he was in a pasture belonging to a rancher friend located on Lee's Creek about ten miles west of Cardston, where he caught two pack horses and a saddle horse left there the previous fall. Without pausing for breakfast, he saddled up and headed back through the thick timber on the head of Lee's Creek over the divide into Belly River. He had won the race, for he loaded up his horses with packs of beaver skins and slipped back over the line into Canada before the American Park Rangers could get to him. They had nothing to offer the Mounted Police by way of evidence, and the Queen's warriors did not get in Joe's way. Late that night he launched a canoe in the boiling flood of the St. Mary's River and paddled east all night in the dark. Taking no chances, he camped by day and traveled by night till he reached Saskatchewan. In due course he arrived eight hundred miles east in Winnipeg, where he sold his hard-earned catch and came back in style by train to southwest Alberta.

It was enough for Joe to win such a race. It was not his way to brag. He grinned quietly and shrugged when asked about his great run, and said very little of anything about it. But the story got out, one way and another, and proceeded to go the rounds of campfires up and down the mountains. Joe told the story of his adventure to the friend who loaned him the canoe in Waterton, and years later, this man told it to me. The campfire version was not embellished by exaggeration. Even the greatest storyteller found no reason to dress up that one. Joe could afford to grin quietly and shrug when asked about it, for he had performed a feat of endurance that has rarely been surpassed.

About 1919, the provincial government of British Columbia launched a campaign against the imaginary ravages of the cougar on big game, domestic livestock, and human beings, and enlarged on the stupidity by declaring a bounty of forty dollars per cat scalp turned in by citizens of that province.

Joe heard about it, and was delighted to know that a furbearer of no previous value was now a real financial potential. He laid his plans accordingly and that fall packed a grubstake into a cabin up on the North Fork of the Flathead River, which he used as a headquarters for his trapline. When the first deep snow hit the high peaks of that country, dressing the trees and slopes in white, Joe and his partner were ready.

His partner was Charlie Nixon, whom I knew well for many years, then only fifteen years old, but tough and ambitious. He was no stranger to wild country, for he lived all his life on the east slopes of the Rockies, where the rigors of a small homestead ranch had taught him the virtues of knowing how to stay alive when the going was rough. Just the same, Charlie told me years later that his first weeks on the trail with Joe Cosley almost did him in. Not only was Joe a merciless traveler on his long snowshoes, but he virtually lived off the land as he went. His grubstake consisted of flour, salt pork, tea, and prunes along with some salt and pepper for seasoning. This was their basic food, but used sparingly. They ate venison, and about everything that got caught in the traps was also potential stew material. To Joe it was a great waste to throw away meat of any kind. Charlie remarked that he got to like lynx and cougar meat, but that marten, mink, and foxes were pretty strong fare. Two or three times during the first few weeks on the trapline Charlie was about ready to revolt at the prospect of eating more of the weasel and fox clan; but then the traps would miss catching very much for a day or two and, remarkably, weasel and fox meat tasted better and better as a result.

When the snow got deeper Joe began taking cougars. His method was unique and very effective.

The deer population of the lower valleys was at a high level and cougars were plentiful. When they crossed a fresh track, Joe took off in pursuit and simply trailed the cougar on snowshoes till the big cat got weary of bucking loose snow and climbed a tree to rest. Then Joe shot it with his pistol and added another scalp to his catch.

One day Joe was trying to unravel a maze of cougar tracks that wove all over an avalanche path on Sage Creek. He was up on top of an old log jam left at the foot of the slide track when suddenly the whole thing caved in under his feet and he fell into a hole like a small cellar. He stood up unhurt, wondering how he was going to get out of the place, and then became aware that he had company. With a great snort a giant grizzly sorted himself out of a mess of dead sticks and snow to stand up for a look at what had come through the roof of his den. Before the big animal could get the sleep out of his eyes and organized, Joe coolly pulled his pistol, rested the fist holding it over the crook of his left elbow, and shot the bear under the chin. It fell dead in a heap with a broken neck.

About that time Charlie arrived to help skin the bear. They dined on some of its meat for supper and later they went to bed in a rude lean-to in front of their fire with the raw bearskin over their blankets to keep out the cold.

Years later Charlie told me, "That was the coldest night of the winter. It must have been about thirty below. We were sure glad to have that hide—lousy as it was. That bear was godfather to a whole tribe of the biggest red fleas I ever saw! You could have hitched up two of them and just about skidded logs with them! Next morning we were all over bumps and itchy as hell, but we had slept warm."

Up and down the Flathead River they ranged all winter, harvesting small furs and running every cougar they could find. After the first month of toughening, Charlie could hold his own, and grew fond of cougar meat. It is light-colored, close in the grain, and tasty. However, like the wintering deer, it is short on fat, and this they craved almost constantly. The bear had provided a good supply and they had saved all they could pack out to the cabin, but when this ran out they depended on the salt pork. Along toward spring, the pork was getting low, and both trappers were feeling the need for fat very keenly.

One day Joe dug down through six feet of snow and set a trap under the ice of a beaver dam. Next morning they had a big fat beaver. They ate almost all of it except the hide. Joe celebrated by running down two cougars that day.

# The Ashes of Old Fires

When spring break-up finally called a finish to the long winter, they had accumulated many cougar scalps, which, along with the other furs gathered, made a fine catch. To get it they had ranged through a piece of wilderness about sixty miles long and thirty miles wide sleeping where night found them and living off whatever the country could offer.

As Charlie said, "We about had muscles built up on our shin bones from running on snowshoes, and we got used to having our bellies growl at us for grub."

The last time I saw Joe Cosley he was still slim and straight as an arrow, and he walked with a long, loose stride that could eat up the miles. When the country here got too tame for him, he moved north to take up a trapline on a creek southwest of the Great Slave Lake on the delta country of the Peace River, where the last herds of the great wood buffalo still roamed in hundreds. Every once in a while one of his old friends down in the Waterton country would get a letter written in a beautiful hand telling about his adventures in the sub-Arctic bush country.

He was over eighty years when the last letter came. We never heard from him again, for away in a wilderness trapline cabin a great mountain man went out on his final adventure all alone. There is a big old aspen on my ranch bearing the inscription, "Joe Cosley, April 18, 1922." Along with it is his insignia, a lonesome heart scratched on the bark of the tree—a fitting epitaph for a man who spent his life in solitary wandering of the places where rivers roared and great peaks cleaned the sky.

3

# Horses and Horsemen

WHEN I WAS A YOUNGSTER GROWING UP ON THE EAST SLOPE
of the Rockies where Drywood Creek cuts down through
the hills to join Yarrow Creek and Waterton River before
making its contribution to the Saskatchewan, almost everything we did was done by our ability to work with horses.
We rode horses to work stock and drove horses in harness to
put up hay and haul supplies. Horses were a part of living on
the frontier, besides being a means of having fun and a mark
of prestige.

Probably no other living thing has had such an impact on our lives as the horse. Likely no other animal has exerted such an influence on the cultural and social development of man. Certainly the opening of the western frontiers of America was made possible by horses that were ridden and packed under saddle and those that worked in harness.

The equine influence reaches far back, to where the first captive horse was likely found in a bog by hunters somewhere in Europe or Asia. Or perhaps it was picked up as an orphan colt to be taken back to the camp alive, raised around the door of the cave and kept as a beast of burden. Maybe early man, tired of wearing down his callused feet on hard trails, watched, as I have done with a sudden lifting of the heart, a wild horse herd thunder by with tails and manes flying in the wind. Covetous of the speed and freedom promised by association with this grass eater, he may have woven a rope from tough strands of grass, or perhaps fashioned one from strips of green hide, and having tied a noose on one end, he hung it on a limb over a trail leading to water and hid to wait.

Who knows what things he caught before a horse put its head in the loop. That first one was likely a mare, for the mares are always in the lead, with the herd stallion bringing up the rear. One can picture that primitive man staring in wonder as his captive pawed the air, squealing and fighting to free itself of the rope. Maybe the horse choked itself into submission, or perhaps the man was crafty enough to use another rope to tie up its feet and bring it under his control. However it was done, it is quite within the realm of reason to suppose he likely collected a fair assortment of horse tracks on various parts of his anatomy in the process. Before he finally straddled his captive's back, thus becoming the envy of every male who came to look, he probably had even more scars to prove the price he was paying for progress.

But suddenly he found that the size of his territory had broadened. Now he could easily journey to distant corners of the country—places he may have never seen before—and cause utter consternation in a strange village, maybe steal a woman, and far outdistance his pursuers on the way back to his own fire pit. Before this such raiding had been a risky business and very hard work besides. Now his movements were freed of much exertion; he could ride a long way and arrive still fresh enough to fight well. He was, for the first time in the history of mankind, independent of the limitations of his own two feet. He had become a personage of note in the eyes of his fellows—a very rich individual.

Then one warm night in spring, when the leaves were whispering among the trees on a soft breeze and the night birds sang, the musky, penetrating scent of the tethered mare carried on the wind to the questing nose of a stallion. And the strong magnetism of procreation drew him in close under cover of the velvety darkness, his muscles quivering at the smell of her. Unable to drive her away, he covered her where she stood, before her erstwhile captor could rouse himself from his sleeping skins and investigate the cause of the commotion. In due course what had been a one-horse sub-chief became a two-horse chief. The man's fortune had doubled within a few moons. He was now more than just a tamer of wild horses—he was a sort of hairy tycoon. His world had shrunk and grown infinitely bigger for the same reason. The location of his calluses had shifted a bit, from his feet to his rump. He was now a horseman.

Horsemen are a very special breed, for the acquisition and use of horses have always added a certain portion of healthy arrogance and pride to the character of man. Horses have also contributed greatly to the development of man's intelligence, for with horses to help him gather a living, his way of life became much easier. No longer did he find it necessary to work himself to the bone in pursuit of game to keep the stew pot full. His hunting was made much simpler, for the animals he preyed on took longer to realize the danger of his approach when hearing the feet of what had always been a harmless four-footed one coming through the forest. Upon jumping a herd of bison or wild oxen, the hunter could now charge in close with ease and lance a fat animal of his choice. If he missed the first try, he could stay with the fleeing animals for a second try or even a third. Perhaps the man's great surge of savage joy upon making a kill infected the wild heart of his mount, which made her something more than just a half-tamed beast—a part of his life.

The horses used in the settlement of the American west were more than just beasts of burden. They were an absolute necessity, without which the frontiersman would have been afoot and the opening of the west likely delayed by a half century at least. Worked in the harness pulling stage coaches, carriages, and various types of freight wagons, the horse was a chief means of power for wheeled conveyances. And innumerable times the early trappers, traders, and cowboys found themselves in a tight spot, where their lives depended on the speed of their mounts. Many are the stories passed down through the years of wild rides through the roughest kind of country, when the penalty for a slow horse was the loss of a man's scalp, the price he paid for losing the race. Under such circumstances a man could easily become an enthusiast for fast horses.

Charlie Russell, the famous raconteur and western artist,

told the story of the cowboy, who, for reasons best known
to himself, was on the move to far pastures with a posse of
grim riders in pursuit. He had outdistanced his pursuers
and was enjoying the view of the near side of a getaway
when his mount threw a shoe and went lame. This put him
in a desperate situation, for a lame horse is not much better
than being afoot, and under the circumstances he could
feel the tightening of the noose on his neck.

Resting his horse, he was standing with the bridle reins
looped over his arm on the rim of a valley, rolling a smoke,
when he spotted salvation coming down a steep draw across
the valley, heading for water. It was a bunch of wild horses,
and among them was a magnificent sorrel gelding, deep of
barrel, clean of limb, and in top condition. The draw they
were following came out on a flat bar by the river just a bit
below a point of a cutbank jutting out toward the water. By
some careful and quick maneuvering, the cowboy rode his
jaded horse into a bit of cover downwind from the mouth
of the draw as close as he dared get.

With his cinch tight and his loop ready, the man waited
for the horses to come down the trail onto the gravel bar. An
old mare in the lead of the bunch trotted out to the edge of
the water, and after a long look around put her head down to
drink. The rest of the bunch lined up on each side of her,
filling up. At this moment the cowboy made his play. Using
every last ounce of his horse's strength and speed, he jumped
in close, and in the momentary confusion as the wild ones
wheeled and milled against the cutbank, the gelding reared
high against another horse, and the cowboy's loop sailed out
as straight and true as the strike of a snake to close around
its neck.

For a moment or two the big gelding struck at the rope,

squealing and fighting; but then he came up short facing the rider, his neck arched, fire in his eyes and ominous snorts rolling in his nostrils.

"When I heard them rollers in his nose, and seen the way his ears was cocked," the man remarked later, "I knew this bronc had seen a man before. When I seen all the brands he was wearin', I knew it for sure. He looked like a map of Montana and Wyoming with all them marks on him. It was a sure bet all the people that had owned him wanted to get rid of him for some good reason. So I begin to get fixed for some high ridin'!"

Untying the hair McCarthy rope off his hackamore, the cowboy threw a loop around the big sorrel's front feet. Right away the horse quieted down, and allowed the hackamore and bridle to be slipped over his head. Then came the saddle, and as it was cinched down the horse stood with his back humped up and his near ear drooping—a sure giveaway that he was getting set for a fight.

About this time the cowboy spotted movement on the hills above, and a moment later a bunch of mounted Indians came pouring down the slope as fast as their ponies could run. What they had in mind was not good, for they were all painted up and rigged for war. It was time to be moving.

The cowboy reached down and untied his hobble rope. Then in almost the same motion he grabbed the cheek strap of his bridle and hauled the sorrel's head around toward him, booted the near stirrup, and stepped up in the saddle. What happened next was the reason for all the brands. With a bawl like a wounded grizzly bear, that big horse put his head between his front legs and left the ground in a great stiff-legged jump. He came down in a jolting explosion of dust and went corkscrewing out along the flat by the river in a string of wild spine-snapping jumps.

A rider less inspired might have been thrown, but guns were going off somewhere behind and bullets were whispering by wide of their mark.

"I wasn't too worried about them savages," the man opined years later. "The way I was moving around they sure as hell didn't stand much chance of hitting me with a rifle slug. That big horse was sure doin' his best to break me in two and kick me out of the saddle at the same time. I lost a stirrup, my sixshooter and my hat. I pulled about everything that would come loose off him and my saddle from his ears back to his tail. But with all that shootin' and ki-yiin' in my ears, you couldn't have chopped me out of that saddle with an ax!"

Maybe a bullet kicked some dirt in his face. Maybe the horse realized his rider was going to stay. Suddenly he picked up his head and began to run. He could run even better than he could buck and in a couple of jumps the scenery was going by that cowboy in a blur. The singing of the wind was music in the man's ears, as the horse tore a hole through it like a spooked antelope heading for far horizons. The Indians pulled up their mounts and watched him go with envy, knowing they had no chance to stay in rifle range.

Without his hat and his gun, the rider felt half naked and a bit helpless in the face of the elements and the people who might be looking for him. But the sorrel carried him far to safety. To his dying day that cowboy claimed he owed his life to that big free-roaming outlaw horse.

BECAUSE RANGE WORK did not allow much time for preliminary conditioning of a horse to accustom him to a rider, breaking methods were cruder and rougher than they are today. Most horses ran free on the range till five years old before being coralled, halter-broke, and then ridden by the bronc-buster of the ranch crew. Once started in their training for working cattle and becoming accustomed to rein, the broncs were turned over to the working crew for additional education through everyday use in working the stock. There were times when such horses were broke to ride before they were trained to lead on halter.

My friend and neighbor, John Wellman, tells of getting a job with a crew trailing a bunch of loose horses from eastern Washington into Montana when he was a youth in his teens. In those days nobody had as yet seen a motorcycle, and the young man of the day projected his image by claiming his prowess as a horseman. In this ride John had lots of opportunities.

The two hundred-odd horses they were trailing were mostly raw range broncs that had never had a rope over their heads. There were a few of them broke to ride, but not enough to keep the crew mounted for such a trip. So they broke horses to ride as they went. They were moving the bunch forty to fifty miles a day, and for the first few days it took some hard riding to keep the horses together in the rough breaks and steep valley country. John, being the ambitious kid of the crew, got plenty of chances to start building a reputation. As a matter of fact, he claimed that the whole crew from the foreman down were so solicitous and helpful he figured they were going to end up killing him. Every time his mount began to look leg weary, which was two or three times a day, they would throw a loop around a loose horse's front feet and throw him. Then they would help John transfer his saddle to the fresh bronc by poking the cinch under its belly with a stick as it lay on the ground. When the cinch had been tightened up and John was in the saddle, they let the bronc get back on its feet.

This first ride was not as rough as might be thought, for seeing the bunch running ahead and being confused, the bronc usually just lit out at a run to catch up. So John would find himself on a fresh horse running in the middle of the bunch. By the time his new mount began to take notice of what it was carrying on its back, it was beginning to tire a bit and answer the pull of the hackamore shank. When the horse began to show signs of lagging, his fellow riders would rope out a fresh one for him. The next time these horses were ridden, it was by the best riders of the crew, for after a rest and a chance to think things over, such horses would generally buck hard. So the trail drive went, and by the time they reached Montana most of the string were well on their way to being good saddle stock. John opined that they probably had more horses broke to ride that knew nothing about leading on halter than any other outfit in the country. In addition, the kid of the outfit had found out that a bronc rider's reputation is made by where he has been sitting and not very much by what he might have said.


UNLIKE RANGES FARTHER SOUTH, Alberta was mostly horse country; there were few mules. The long-eared hybrids were for the most part distrusted by northwestern horsemen. Mules, largely a cross between a male donkey and a mare, have many of the better qualities of both species, but they can also display a combination of both kinds of hellraising too. A mule can carry a grudge for a long time. It can be as gentle as a pet dog with someone it knows, and, sometimes, meaner than a scalded cat with a stranger.

My experience with mules has been limited but salutary,

and while this is an account concerned with horses, for comparison's sake there are a couple of adventures with mules that deserve telling.

A bachelor who lived on a small ranch a few miles from us had a team of small mules he used to pull a light wagon. They were as independent as animals can get and had about as much respect for a wire fence as elk. Consequently they were well known to everyone in the country, since they regularly took off on a tour of the neighborhood. They were particularly unpopular with the local housewives, who took a dim view of their sampling of vegetable gardens.

Forty miles farther north another rancher had a team of huge mules, which he used for many years for all kinds of heavy draft work. Both were becoming old when one contracted a bad fistula and died. A lone mule is not much good to anybody; and the survivor of this team was no exception, for it never ceased looking for its mate. It became a sort of wandering long-eared hermit of the hills, disappearing for months at a time and then reappearing in the most unexpected places.

Many were the stories that went around concerning this mule. I had never seen it, and like a lot of people considered it more or less a sort of legendary myth. Then one hot afternoon I was riding in the dust kicked up by fifty loose horses on the short end of a two-day drive to a base camp in the Oldman River up on the flats where Racehorse Creek joins the North Fork. I was making this drive alone, which kept me busy during the first day, when some of the bunch kept trying to find a way to get back to the home ranch. Fifty horses running free can make things interesting for a lone rider trailing them, but I was well mounted and had not lost one horse.

I was jogging along feeling tired but satisfied at pulling

off a tough job. I was on the last fifteen miles of trail, going through the Livingstone Range at the Gap. It was blazing hot and there were streaks of muddy sweat showing dark on the dusty hides of the horses strung out ahead. Like me they were getting leg weary and anxious to get to camp. Most of the bunch had been there many times, a jumping-off point for pack trips through the mountains. They were reeling off the miles at a steady trot with the jingling of the Swiss bells of the pack horses accompanying their going.

Over the sound of the bells there came a noise like nothing I had ever heard before. It had an eerie, unreal quality about it and seemed to come from nowhere in particular. My saddle horse, a bay gelding of somewhat nervous disposition, heard it too, twitched up his ears, and swung his head around to look up toward the top of a timbered ridge flanking the trail. Then the sound came again, a sort of awful hoarse scream ending in a coughing roar. It was unlike anything one would expect to hear on a sunny afternoon among the peaceful Alberta foothills. Had it been dark the sound would have made a man's hair stand straight up, for it had a quality like something on the near slope of a very bad dream.

Then out of a fold in the hillside came the source of it, going at a high gallop—a monstrous bony old mule with a head like a wheelbarrow and a ridiculous string of a tail flowing in the breeze behind him. He had spotted my horse herd going up the valley and was bent on coming down off the hills for a visit, maybe hoping he would find his long-lost mate running with them. About every fourth jump he opened his great mouth and let out the most god-awful bray imaginable. He would have won no prizes in any kind of music festival.

By this time my saddle horse was fixing to throw a fit,

and before I realized what was happening he let out a sharp snort and bolted straight ahead through the bunch. That was all it took to spook my horses, and in two bounds every one of them was in full stampede. There were rocks as big as eggs and chunks of sod flying past me as those horses humped up and dug for toeholds to get out of there. My horse was fast, and before we had gone a quarter mile we were up with the leaders, burning a hole in the wind. The old mule veered in a swing along the hill to cut us off, but the horses did not give him much chance. He was braying steady, trying to talk us all into stopping for a pow-wow, but none of the bunch understood his language. Those horses were never more inspired to run. They just left him as though he was tied—bawling and braying in their dust. When the procession slowed a bit, a mile or so up the valley, I eased back along one flank of the bunch to see if any were missing. They were all with me, but now their hides shone with sweat and their nostrils flared wide in their reach for more air. But the weariness was forgotten. We came into camp in record time, stepping lively. I had learned that range horses have a tendency not to recognize their relation on occasion—especially when they have never seen him before.

I never had the chance to see mules handled and worked till I went down into the Salmon River Wilderness country of southern Idaho early one spring as special guide and photographer for a party from Chicago.

Bill Sullivan was our outfitter, a genial Irishman with the heart of a poet. He was a fascinating character, with a wealth of experience on the mountains, and he could make up poetry on the spot about most anything that happened on the trail. On occasion his rhymes were completely unprintable, but always amusing, and we found him hilarious company around the evening campfires. He used both horses and mules in his pack string.

He had a mule—or more correctly what is known as a hinny—in his outfit, a somewhat unusual cross between a stallion and a jenny. The get of this kind of mixup is usually smaller than the average mule, by and large maybe a shade more intelligent, and perhaps capable of more refinements of pure hellery. This one went by the very unoriginal name of Jenny.

Jenny was a pretty smoky-blue color, her proportions were good, her features clean-cut and delicate, her legs trim and neat, and her eyes had a deceptively friendly look about them.

We packed up at a place called Meyer's Cove, in a big basin on the pass at the head of Camas Creek, which heads down through the mountains westward to join the Middle Fork of the Salmon River in the heart of some of the wildest, most rugged, and beautiful country I have ever seen. It was a sunny morning in mid-April.

When Bill led in his little blue mule to pack her first, I offered to help, but he shook his head. "She's a one-man mule, this one," he told me. "Sure as hell she'll throw a fit if you touch her. She looks real gentle, but she's a faithless bitch for sure. Best not go near her, for she kicks like a streak of light and if she misses she didn't mean to hit you!"

I watched her pack being put on and lashed down with the diamond hitch, and it was hard to believe what Bill had said, for she never made a wrong move. But when he led her away, she sidled around trying to get within kicking range of Bill's English sheep dog, and when he jerked her back, she flashed him a nasty look.

Down in that part of Idaho packed mules and horses do not run free on the trail when moving from one camp to another as they do in the Canadian Rockies. Each wrangler and guide leads a string of four or five animals tied together. There is a loop of light rope knotted through the rear fork of each pack saddle and to this is tied the halter shank of the pack animal coming behind. This loop is called a "breaking string" and in event of trouble it is designed to break so that halters will not get broken or an animal dragged.

This morning as we strung the pack outfit out along the trail winding down Camas Creek, Bill was in the lead with four horses and the gray mule tied at the rear. I was next, and the cook was behind me with his string. The party from Chicago was bringing up the rear.

Being directly back of Jenny, I had a good opportunity to observe her, and because mules were more or less new to me I was watching her close. For the most part she trailed along as sedately as a Sunday-school teacher walking down Elm Street on the way to administer the good word to her class. But every once in a while Bill's dog would come gamboling up alongside the string, and when this happened, Jenny peeled a wicked eye in his direction as though judging the range for possible murder. The dog seemed aware of her intentions, for he stayed well out of reach.

Things proceeded smoothly for several miles. Then the trail dropped into a twisted gorge, where for the most part it was graded out of the loose rock flanking the creek on both sides. Here the mountains were high and jagged on both sides of the canyon—so high in fact that the water-eroded walls above us hid their tops, and we were only aware of being overhung with peaks because of more distant vistas earlier. Winter snow was melting off the high slopes, and all the little side canyons were running white water. The gorge was full of the sound of water leaping and plunging toward lower ground.

Even at these lower levels snow still lingered in some of the shadier places. Some time after noon we came to a spot where the trail forded a small, almost perpendicular gorge dropping off the mountainside. On the high side of the trail, a stretch of solid ice canopied the boisterous little stream. It was steep as a church roof and slippery as glass. Bill's string was about halfway across this little ford when the dog chose to come running along the slope above us heading for his favorite spot just ahead of Bill's horse. When he came to the ice, he suddenly lost his footing and came sliding down toward the trail. In a flash the little gray mule slammed into the horse ahead of her and pushed him far enough to put her into position to get at the dog. Although I saw instantly what she was about to do, she had kicked the dog hard three times with unbelievable rapidity before I could jump my horse up against her rump and shove her ahead. She kicked my horse violently with both hind feet as she went, and she would have probably gotten me in the shin with the next one had not my sudden yell shaken her a bit off balance, enough to make her hesitate. Then the slack went out of her lead rope, jerking her ahead.

The dog lay in the stream inert and unconscious. When I dismounted to drag him out, it was like handling a hunk of sodden wet blanket. At first I thought he was dead, but he let out a sort of retching gasp for air and began coming to life. As far as I could tell, no bones were broken, so I laid him out on top of a flat boulder in the sun to recuperate, and we rode off and left him there.

He caught up to us that evening when we stopped to make camp. He was about the sorriest looking dog imaginable—there being fewer more down-hearted looking things than a sad English sheep dog even without a working over by a mule. He acted as though he was lame in all four legs and was unable to make up his mind how to limp on all of them at once.

While we were cooking supper, Bill told me how things got started between the dog and the mule. About a month earlier Bill had been offered Jenny at a very reasonable price by a neighboring rancher. He looked her over very carefully, trying to find out what was wrong with her, for the price did not follow the usual trend in mule deals. But he decided to buy her anyway, for she appeared to be sound and young.

After paying the man, Bill remarked, "Seeing as how I seem to have bought a hinny and seeing as how you no longer have any good reason to lie about her, what is wrong with her?"

"Not a thing," replied the man, "except that she won't work!"

When Bill got his new acquisition home and packed her the first time, he found out what his neighbor meant. Just as soon as he tried to lead Jenny, she lay down, rolled her eyes up into her skull, and refused to get up. Thinking maybe the pack or the saddle was pinching her somewhere, Bill took everything off her, whereupon she leaped to her feet, kicked a pannier halfway across the corral, and tried to bite him. Bill patiently packed her again, and again she lay down, the picture of utter dejection. So Bill tied her feet together solid so she couldn't get up even if she wanted to. Then he stretched out in a shady spot beneath a tree to watch and cogitate this problem wrapped in silky gray hide. After about an hour Jenny's ears began to twitch, as though she was trying to figure out this new development. Bill saw her rump muscles bunch up as she tried the foot ropes. But when he went close to her, she became inert again, as though the thought of getting to her feet with that load still tied to her back was the last thing she ever intended to do.

About this time the dog came trotting into the corral and gave Bill an idea. Leaving Jenny tied down, he set the dog on her. It was not every day that the dog had a mule so completely at his mercy, and he waxed extremely enthusiastic about this chore, working himself up into a high pitch of excitement, leaping all over her, nipping lightly here and there and barking in her ears. After a short session of this it was obvious Jenny wanted no more of it, so Bill slipped the tie ropes off her feet. Instantly she leaped up and chased the dog out of the corral. From that moment on, Jenny was a working hinny. She also was a bitter enemy of the dog and never passed up a chance to stalk him. She caught him finally in the little creek, and only by pure luck failed to kill him.

What surprised us was the fact that from then on she chose to ignore the dog. It was as though she figured the score was even, for she never again went out of her way to kick at him. The dog avoided her like poison.

Jenny intrigued me, for I had never seen a pack animal quite like her. She performed as though she took satisfaction in her work, quietly, unobtrusively, and yet with a certain pride, in complete contrast to her earlier balky refusal to carry a load. She treated her pack as though it was made of eggshells, and knew within half an inch the clearance required to get it through between two trees. If there was insufficient space, she would stop and shuffle around through the brush till she found a way that suited her. Never once did I see her so much as ruffle the pack mantle by scraping it on anything.

As Bill told me, she was very shy of strangers, especially when running loose, and she wouldn't let me anywhere near her. Occasionally, when she was tied up, I went in close to rub her back—something she obviously enjoyed. One morning after we had been on the trail nearly two weeks, I took a halter and walked into a clump of sagebrush where she was hiding. Coming up to her slowly and easily without hurrying or being too direct about it, I slipped the halter over her head and led her out to where Bill and the cook were catching the rest of the string.

Bill saw me coming, tipped his hat back, and exclaimed, "Sure now would you look at that! You must be gettin' to smell like a Salmon River savage!"

Whether or not this was a compliment I was at a loss to know, but at least I enjoyed the satisfaction of having struck up a friendship with a wise and independent little mule.

Riding was so much a part of our lives that most of us grew up unable to remember when we first straddled a horse. A lot of the youngsters born on the prairie and along the mountains fifty-odd years ago rode on the front of a saddle with their parents before they could walk. We literally grew up on horses and some of us even learned to think like horses. Certainly most of the time we smelled like them, for we mostly rode bareback till we were old enough to work on the range, when we acquired boots and saddles.

I remember standing one day on the main street of a little town by the Salmon River in Idaho when a small pack outfit of mules and horses showed up, coming down off the surrounding mountains. A young man all dolled up in his best finery rode in the lead with a string of four horses. Behind him came his wife leading a couple of mules. When they were just about opposite me in the middle of the street, the pack mantle on one of the mules jerked a bit, and a tow-headed little boy stuck his head up to announce excitedly, "Hey! We're in town!" Next thing, a little girl showed up out of the opposite pannier and both kids began pointing and jabbering. Apparently they had fallen asleep during the long ride and had just wakened up. In that country, the chances were good that was the family's only means of transportation, since much of Idaho is too steep for wagons or four-wheel drive trucks.

When my brother and I were young, the country schools along the mountains—one-room affairs scattered here and there among the poplar bluffs—all had barns and hitch rails as a part of their regular installations. Drywood School, where we learned the rudiments of the three Rs, was for some unknown reason built on top of a hill overlooking the creek valley and a vast sweep of mountains. The view was unsurpassed, but by long odds it was also the windiest, most unsheltered place in southern Alberta. The playground was excellent for tobogganing and skiing, but just about impossible for playing ball. A batter could knock a ball so far off that butte that it was pure luck if we ever saw it again. What was perhaps most unusual about the place was that the yard and barn were usually full of horses. When I was in first grade, there were twenty-two students attending classes with eighteen horses as their means of transport. Because some of these youngsters had not been within reach of a school when younger, I shared first grade with two boys about twice my age. We all had our own saddle horses. Some of the kids had a team and buggy or cutter, depending on the time of year.

When final bell rang in the afternoon horses poured off that hill in all directions. In fall and spring horse races were daily events in good weather. Once in a while, when the spring grass was lush and somebody was riding a colt or a saddle horse fresh off the range, we would be treated to a bucking match, which usually ended up with the would-be bronc-buster getting "planted" to the accompaniment of delighted howls from the audience. Horsemanship was something we learned more or less unconsciously—a sort of free benefit sidelining more prosaic studies like arithmetic and geography, and one we accepted with vastly more enthusiasm.

Once in a while someone would get shook up in a fall to the point of looking a bit wan and gimpy for a day or two, but broken bones were rare. We were not exactly accident prone, although sometimes recklessness caught up with us.

A friend named George and I caught a ride home one afternoon with a girl who drove herself and her little sister to school with a team. She was about the most reckless, devil-may-care driver I have ever seen, and her usual method of getting off that hill was a fair sample. She would line her team up with the open yard gate, bat them across the rumps with the whip, and as they jumped into a run she would throw her head back and warwhoop like a Commanche on the warpath.

This time George and I stood up in the back of the buggy hanging onto the back of the seat with our toes hooked under an iron brace bar to hold us down when the buggy hit the rough spots. It was a balmy June afternoon, the last school day of the year, and we were all feeling wonderful as we contemplated two months of summer holidays. The horses were feeling their oats and so was Ruth. We came off that hill as though we had been shot at and narrowly missed. We hung on desperately to anything that looked like a decent handhold as the buggy and runaway team came careening down into the valley. There Ruth played the whip on them and lined them up the trail; they went flat out with their tails and manes streaming in the wind. It was wild and reckless—we howled like savages and gave not a thought to a rough ford across the creek ahead. The horses hit the stream at a gallop and water went feet over their backs as the buggy lurched and jolted over the stones. George lost his grip on the seat and fell on the back of his head with his toes still hooked under the brace bar. Naturally he was dragged through the creek with his head bouncing over the rocks until we were a hundred yards away from the other side. There we got down and solemnly gathered around to contemplate George, who looked like the next thing to being dead.

He was very still, his face was covered with dirt and some blood that was running out of a cut on his forehead. Feeling a bit scared and unsure what to do next, we unhooked his toes and gently lowered him to the ground. Then I got a hat full of water out of the creek and sluiced it over his face. The shock of the icy stuff made him stir a little and then he began coming back to life. Before long he was able to crawl up onto the buggy seat with some assistance. Apart from a short lapse of memory and a slight concussion he was good as new in a couple of days. His good solid Swedish skull had served him well.

During the summer we almost lived on horseback. There wasn't a kid in the country that didn't have one or two horses for his or her own use. These carried us uncounted miles as we rambled the wilds exploring, fishing, visiting back and forth, and doing ranch chores. Most of us rode bareback, using saddles only when doing something that required them. When a horse we were using began to look a bit jaded from constant work, we caught a fresh one. It was a wonderful carefree kind of life, and horses were a large part of it.

Like every kid raised on a ranch, where lariats are in almost daily use, we had ropes with which we practiced on about everything. I remember one old rancher remarking that his boys roped about everything that stuck up off the ground, and that "even the damn fence posts duck when they ride by." We roped chickens, geese, dogs, cats, calves; and sometimes each other.

Once I was all dressed up on a Sunday afternoon waiting for some guests, wandering morosely around with nothing to do and feeling mighty uncomfortable in the unaccustomed finery. Dad's saddle rope caught my eye. He did not look kindly on anyone using his rope, but I took it anyway and began throwing loops at the end of a rail sticking out from a corner of the corral back of the barn. Then a big steer came to lick salt with the milk cows not far away. With no real intention of catching him, I sneaked up and threw a loop at him anyway, more to spook him than anything else. He threw his head up and caught it beautifully right around his horns.

He wheeled immediately and headed for the timber. For want of something better to do, I hung on. In two jumps that steer was at a full run with me trailing like a wet saddle blanket on the end of the rope. The ride didn't last very long, for something jolted me in the belly hard enough to knock my wind out and loosen my death grip on the rope. My hands got burned as it shot from my grasp. When I got to my feet to survey the situation, I found myself reeking from head to foot with fresh cow manure and stinkweed. Smelling to heaven, I sadly inspected my rope-burned hands and decided that it was a whole lot easier sometimes to rope something than turn it loose.

Dad had been watching the whole adventure. When he appeared I got a real hide-peeling chewing out. When angry, he could exercise words that would blister the skin off the toughest. He knew how to make a point and have us boys take notice. When he was finished he saddled his horse and headed out after the steer to retrieve his lariat. As he went

I noticed his grim expression had softened and he might even have been grinning a bit although he was trying to hide it.

Heading for the house to collect another scolding from my mother, I came to the conclusion that fooling around with a lariat could sometimes be about as dangerous as playing with a loaded gun.


LIKE THE LESSON of the roping episode, the lessons learned in range work often came unannounced and completely unexpected, and they were the kind that lasted for a lifetime.

I was still pretty young, maybe about ten or twelve, when Dad and I had a run-in with a range bull that was the kind of milestone on the road of experience never forgotten. One of our bulls went visiting a neighbor's herd, where he was making a nuisance of himself fighting their herd sires. So we rode out to bring him back.

We cut him out of their bunch without any trouble, but the closer he got to home, the more he tried to break back. When we got him to our pasture gate, he suddenly wheeled and charged my horse. My mount jumped out of his way and the bull headed for the timber at a dead run. Like a flash Dad's rope was down and about two jumps from the trees a loop shot out to snap shut around the bull's horns. Dad's big saddle horse tucked his rump down and set his feet to stop the runaway animal, but the off-side latigo holding the cinch broke, whereupon Dad and the saddle shot over the horse's ears. He hung onto his dallies and still had both feet in the stirrups as he flew out of sight into the aspens.

Then all hell broke loose. The timber began to break, the bull was roaring, and I could hear profanity crackling like red-hot hunks of iron hitting cold water. Then out of the general uproar, I heard Father ordering me to come help tie up the bull. Right then I was scared and wanted no part of this mixup, but Dad was in trouble clear over his head and by the sound of him it would be wise to do as he said, so I sneaked over and peeked into the timber from behind a big tree. What I saw did nothing to calm my nerves.

Dad was under his saddle still hanging onto his dallies, and the saddle was jammed into the base of a tree. The bull had run around another tree two or three times and was so mad he was fairly screaming as he tried to get back at us. But he was tied up too short and the rope was new and strong enough to hold him. Under Dad's direction I got hold of the loose end of the lariat and took some wraps on another stout tree. When all was ready Dad slipped the dallies off his saddle horn and I quickly took up the slack.

Dad then took his saddle back out into the open where his horse was waiting. Fashioning a latigo out of a bridle rein, he soon had his saddle back on the horse and cinched down solid. Then he came riding back to take charge of the bull. While I was scrambling to get back on my horse, he and the bull burst into the clearing. The bull promptly charged his horse, but that wise mount neatly sidestepped him and away they went at a dead run with the bull in the lead.

Dad did something few people know how to do any more. He rode up about even with the bull's near hind-quarter, threw some slack rope over his back onto the off-side and then rode off at an angle away from him to the side at top speed. When the slack came out of the rope with a sudden jar, the bull left the ground completely with all four legs swept from under him, took about a half turn in the air and came down on his side like a wagonload of bricks. The wind was knocked out of him, so he couldn't get up for a while. When he finally got back to his feet, all the fight was out of him.

It was likely that same spring that Dad started me heeling calves for the branding. We were short-handed. I was too light to be much good in the wrestling team, and John was old enough to handle the irons at the fire, so I suddenly found myself promoted to the job that usually goes to the top hand of an outfit. Since I was short in the leg and my feet wouldn't reach the stirrups, I just poked my toes through the stirrup leathers of Dad's roping saddle.

This kind of roping requires flipping a loop on edge under a calf's belly so when it jumps both hind feet go into the loop. The roper then snaps the loop shut with a flip of the wrist, takes his turns on the saddle horn, and heads for the fire dragging the calf behind.

Most of my practicing had been done from the ground, although some of it had been done from a horse on our pokey, gentle milk cow calves, so I missed a lot. But that big horse of Dad's knew his business, helping me every way he could and making me look better at this new game than I really was. I felt about ten feet tall when I dragged in that first calf. When twenty had been roped and branded my arm was getting tired from throwing the rope, but I hardly noticed it. By the end of the morning, my right arm felt as though it was about to fall off, and I was having trouble making my loop behave.

But Dad was patient and let me finish the job. The last calf gave me a bad time. Having been roped at several times, he was getting foxy, and many loops missed him. Finally I got mad and threw the rope on his neck. Dad was leaning against the corral sharpening his knife as he coolly surveyed my last catch skidding in close to the fire stiff-legged with the rope on his neck instead of his heels.

"You better get off and show us how to throw a calf with a rope on that end of him," he said.

So I got off and started down the rope for the calf feeling as though I had put myself out on a limb and was about to saw it off between me and the tree. The calf was wild but not very big. The first thing he did was jump on my foot with a sharp hoof and I wanted to kill him. With a short left-handed hold on the rope by his neck I reached over his back with my right hand and grabbed his flank. He jumped and more out of pure luck than anything else I heaved at exactly the right moment. He went down flat on his side, with me still hanging on with both hands and my knee planted hard on his ribs.

I was feeling a bit mean and raunchy and without even thinking, I said, "Now if somebody will come flatass this sonofabitch, I'll go unsaddle my horse!"

A neighbor who was helping us snorted through his nose as he grabbed the calf by the heels and sat down behind it. Somebody else grabbed a front foot and knelt on its neck. I took my rope off it and was about halfway across the corral, suddenly feeling appalled at being so cheeky, when I sneaked a look back past the horse trailing behind me. Dad was standing with his hat tipped back scratching his head, with a half-surprised, half-amused look on his face. He never said a word.

FOR MANY YEARS we had at least thirty horses running in the mountains back of our ranch. They were wild as hawks—big, fine-looking broncs that knew every fold of ground and twist of the trails for miles. About twice a year we rounded them up to look them over, cut out some to be gentled, trimmed hooves, gelded the stud colts, and branded any that needed to be marked.

After we had ridden enough with Dad to know the country, John and I took over this roundup. We rode top horses, for a sure-footed mount was a must. Any other kind was likely to fall and break your neck in that steep standing-on-end country among the peaks. We held a big advantage in being light, so our horses hardly noticed our weight and could carry us a long way at top speed without getting tired. It was a high, wild kind of riding—the kind one grows to hate or love, depending on his training and whether or not he is afraid.

Although we sometimes got painfully snagged or skinned when we came too close to a tree, we loved it. Once in a while we had a horse fall under us, but came out lucky, with no serious injuries. Half the secret of knowing how to ride is learning how to fall, and this we learned by the simple means of being thrown. Listen to someone brag about how he never comes off except to step down and you will be looking at someone who has not done very much riding or work from the back of a horse.

One spring we were bringing a wild bunch down off the side of Drywood Mountain along the steep-pitched top of a hogback ridge flanking a ravine full of hard snow. The horses ran down to a spot where the draw narrowed to a gut choked with drift, slowed to a deliberate walk, stepped across a streak of hard ice and then hightailed it back up the mountain. John was ahead of me on a fast little mare, and he turned her out across the snow on the run to head them off. Everything went fine till she stepped on a patch of ice and fell sprawling.

They must have slid about a hundred yards before they

73

stopped. At first John was on the low side, but she rolled clear over him. When they stopped, he was sitting on the high side of her trying to get his wind back, with a shovelful of snow inside his shirt. He wasn't even shaken up or scratched.

We had to start all over again, but the next time we hazed the horses down off the mountain we gave them no chance to repeat their getaway. Running wild mountain horses is about the best way there is to learn how to think like one, for unless you could anticipate their moves far enough ahead to block them, they could give you a bad time. If they got away with cutting back, it spoiled them and made them much harder to handle.

We liked best to spot a bunch we wanted from a distance. This gave us time to plan a strategy to fit the lay of the country we had to drive them through. If possible we would get in close before we jumped them and then we rode like the wind as close to their tails as we could get. Ki-yiing like a pair of Blackfoot braves on a horse-stealing raid, we did our best to keep them off balance a bit, never giving them a chance to think until we had run some of the fire out of them. Then we usually had them in more favorable country, where we could handle them easier.

BY THE TIME we had reached our teens, there were few horses that could get away from us. We were both riding the salty ones and not nearly so careful about how we got a mount warmed up on a chilly morning. We had learned a few things about horses and the power of endurance. If a man was determined enough, well mounted, and had sufficient horse savvy, there was just about nothing he couldn't do by way of putting them where he wanted them. If he wanted to bad enough, he could ride about anything, too, but we all ran into limitations sooner or later.

While there is a good deal of truth in the old adage that "there never was a horse that couldn't be rode, and never a cowboy that couldn't be throwed," it is a certain combination that counts when a man undertakes to ride a rough one. When a good rider can find and make use of the sometimes hard-to-see advantages always present, he can quite often make riding a bad horse look easy.

But no matter how many horses a man handles in his lifetime, no two of them will be exactly alike, for like all animals, including humans, they tend to be individuals with certain likes and dislikes, forms of habit, and identifying characteristics. It is these differences that make working with them and handling them so interesting, and sometimes an enigma. Inevitably, when long experience and constant exposure bring a man to a point where he feels he is master of about anything this four-footed friend can present, one comes along that reduces him to a certain level of added respect.

There are, of course, all kinds of horsemen. Today the differences are not so noticeable, since so few people train horses, and only the better ones can make a living at it or draw the attention of others. It has become a clear-cut matter of business or pleasure. But when everyone in rural areas used horses for work or pleasure, comparisons were wide. Some were sadistically cruel with their horses, and these were held in contempt by their neighbors. The cruel man was generally one who was afraid or hated animals—a misfit in the general picture in which he found himself a part.

Sometimes the bronc riders employed by the bigger outfits to ride and work the inevitable collection of bad horses owned by most ranches became unnecessarily callous and

perhaps cruel by being constantly exposed to violence. For
an extra ten dollars above average cowboy pay he started
out each working day by saddling up a horse nobody else
would have anything to do with. He knew the horse would
throw him if it could, and kick him or maybe even bite him
if he gave it the chance on the ground. If he was not riding
outlaws, he was riding raw broncs, and watching one of
these riders made a man wonder if the extra pay was worth
the risk. But money was not all that was involved by any
means. There was a certain prestige that went with the position. A man might otherwise be an illiterate nonentity—a
mere unrecognizable among cogs—but if he was a superlative rider employed to work the rough string, he was next
to the foreman in the social strata of the bunkhouse, and
even if his fellow cowboys might generally hate his guts, he
was respected for his nerve and skill.

Generally a rough-string rider had his collection of old
injuries to remind him of battles lost. On cold mornings
these hurt and stiffened a man, making the saddling and riding of a mean bronc anything but a pleasure. So when the
man stepped up in the middle of his mount and found himself in a wild, spine-snapping, end-swapping, saddle-pound-
ing storm, one couldn't blame him much for feeling like
fighting back with every means he had. Thus it was that
spurs and a heavy shot-loaded quirt came into play in ways
that sometimes drew blood and could be charged to unnecessary cruelty. But having been a rough-string rider and knowing how a man feels under such circumstances, I can say
that there are two sides to the story. Like bulldogs bred and
kept for fighting, some horses seem to take a real joy in
trying to throw a rider, and some of these will use every
means within their reach to do so. There was only one way

to get along with this kind, and that was to show them that life was a lot sunnier and easier if they behaved themselves.

I remember once hearing a rabid member of a humane society ranting and raving about abject cruelty to poor "dumb animals" used as rodeo stock. I got a dirty look when I asked him if he had no concern for the "dumb animals" that rode them. Having been mixed up on occasion in arguments with so-called "poor dumb animals" that were real keen about trying to unload me on the nearest pile of rocks, rub me out of my saddle on a convenient snaggy tree, or just break me in pieces for the fun of it, I have sometimes wondered about the practical working knowledge of humane societies, even if I do subscribe heartily to their general principle. There have been times when I have lain awake in my sleeping bag worrying about what I had to ride when I got up in the morning. At low times like this in life, a man is wont to search his soul for a broader definition of "dumb animal," and perhaps even come to some conclusion about his own possible connection with that general category.

A good number of the horses we worked with could be classed as only semi-domesticated, since most of them were never handled until they were three to five years old. They ran on the open range or in large pastures until being rounded up for breaking to saddle or harness. I use the term "breaking" loosely, for by no means did a good horseman actually break a horse's spirit, and most of the time a bronc-buster was primarily interested in not getting broke himself. Many horses, particularly wilderness packtrain animals, never got to know what it was like to live in a barn or develop a taste for grain. The ground herbage over most of the western range is rich feed, which ripens and cures naturally on the stem in the fall and thus retains a high nutritive quality all

winter, keeping range animals fat and in good condition.
Team and saddle horses kept around a ranch generally were
fed grain while working.

Probably more horses were gentled and trained at the
ages of three and five than four, for horses change their
teeth at four years, a time when their general condition is
low and their mouths are often sore. Because of this the prob-
ability was much greater of producing a "cold" mouth on a
horse, a term to describe a condition wherein a young horse
acquired the habit of being very stiff-necked and difficult
to rein.

One rancher I knew, Frenchy Rivière, who owned many
horses and had a large family, employed a unique system
for breaking his annual crop of young horses. He started his
broncs on their training at a year old.

Come greening-up time, when the range horses have just
shed their winter coats and are beginning to put on fat, he
would round up his herd of range mares, colts, and yearlings.
Fresh off the mountains they were as wild as deer, but when
Frenchy cut out the yearling stock at the corral and turned
the rest loose, it was not long till these young horses were
introduced to training. Each of the children was assigned a
colt to halter break and ride, with only a hackamore. The
colts were ridden bareback, a practice that not only made
good riders of the youngsters, but also brought the colts
into very close contact with them. Colts and youngsters
have a natural built-in affinity for each other, and it is generally amazing how soon a child with some instruction, experience, and knowledge of horses can gentle a wild colt.

As fast as various members of Frenchy's family gentled
a colt, another was assigned, and so it went till the whole
new crop of young horses were broke to lead and ride. Then

they were turned loose to mature before being rounded up again to be ridden and worked.

WHEN I WAS still too young to be capable of breaking a horse to saddle, I helped my father break work horses to harness. Most horses were five years old before being put in harness, for they needed to work with gentle horses of mature age at first. Generally such a horse was broken to lead on halter when younger, so the initial harness training did not include that job.

First my father would rope a bronc by its front feet and throw it. It was then hog-tied and its mane and tail were trimmed. Its hoofs were also carefully pared. Besides facilitating these operations, the tying-down of a big horse usually tended to give the animal the idea that he was vulnerable to the strength of a man, so his initial reaction to being harnessed and hitched to a wagon was likely to be much less violent.

Following the trimming operation, the bronc was let up on its feet, and would then generally show some signs of responding to a man's wishes. Prior to fitting harness and collar, a hind foot was tied up with a rope running from a loose loop fastened around the base of the horse's neck. When this was properly adjusted the horse was required to stand on three feet, and as a consequence was unable to kick. To ensure the habit of standing still, hobbles were usually fastened to the front legs. So as long as the horse stood still, he was largely comfortable, but if he attempted to kick or jump away he was instantly checked and rapidly became aware that he was punishing himself.

When collar and harness were buckled on securely, a heavy ankle strap fitted with a steel ring about two inches in diameter was buckled around each front leg below the fet-lock. A third ring was hung from the belly band or the bottom of the collar. On this combination an arrangement known as a "flying W" or "Spanish hobble" was tied with a long rope. This was simply a trip rope by means of which a man with a single pull could take a bronc's feet from under him. The first driving lesson was usually administered in a corral by means of long reins and this rope.

With hobbles and hind-foot rope removed, the horse was urged to step out. Thus it was introduced to the feel of the bit in its mouth and the rudiments of stop and go commands. If the bronc became unduly active in a bid for supremacy, the trip rope quickly brought him under control. It did not take a horse long to learn what the command to "whoa" meant with the trip rope to accent it. So went the first lesson.

After a few hours' rest, the bronc was usually hitched to a wagon with a well-broken horse for initial instruction in pulling a load and becoming accustomed to the rattle of a vehicle being drawn behind. The bronc's halter shank was usually tied back to the hame of the gentle one, thus forcing it to follow. Generally a horse employed to help gentle broncs was experienced and wise to the job, well able to keep the unruly one in control with ease. Some would even undertake to punish a stubborn bronc if it was unduly rough. Needless to say, setting out in a wagon drawn for the first time by such a combination was full of exciting action.

I recall helping my father hitch up a big leggy Clydesdale mare one bright morning alongside a rangy gelding of the same breeding. She was a tall, Roman-nosed female with lots of fire and action in her makeup. She did not overawe her teammate though, for he eyed her over a bridle blinker coolly, as though remarking, "Behave yourself, or you will land on your bony head!"

When we got everything set and ready, Dad eased up onto the wagon seat. Then I opened the corral gate and took my place beside him with a good grip on the trip rope. Dad spoke softly to the gelding and he eased the wagon ahead, leading the mare at the same time. Upon feeling the collar move on her and touch the single tree on her hocks, she lunged ahead, and the next instant we went through the gate heading for open country like a shot out of a cannon.

"Let her go!" Dad yelled.

With just enough slack taken out of the trip rope to keep it from tangling, I did just that, while he sat back with his boots cocked against the dash, the reins, or "ribbons" as we sometimes called them, firmly grasped in his hands and a cocky, devil-may-care grin on his face.

The wagon rattled and bounced, weaving a little from side to side behind the running team. They were going flat out, but more or less under control, for the gelding knew all about this game and was keeping us out of trouble by outrunning and leading the mare.

My mother heard the commotion and came to the kitchen door just in time to see us go flying across the yard, through the open gate, and skid on two wheels around a bend out onto the trail leading across the ranch.

With the wagon swaying along behind the team like the tail of a kite, we rattled down through the bottom of a shallow draw, bounced up the other side and then shot down along a strip of willows growing alongside a slough. Here Dad swung the gelding in a long arc out across an open meadow, and by the time we completed the big circle, the mare was blowing hard with streaks of sweat showing on her hide. It was easy to pull them down to a walk and head them back toward the corral.

The big gelding was dancing a little, feeling proud, and

showing off his neck in a bow and flashing his white socks. The mare tagged along, lagging a bit at a long walk, mouthing the unfamiliar bit. As we went back into the yard through the gate, a piece of wire sticking out from the gatepost hooked the mare's trace chain close to the single tree, and squealed as it was pulled through a staple. The sound and the sudden jerk spooked her, and she made a great sun-fishing jump as she kicked at it. She took us by surprise and even caught the gelding off-stride. In the wink of an eye we were flying straight for a steep coulee bank, where a big spring flowed down past the house.

Bracing himself, Dad hauled back on the reins and yelled, "Whoa!"

Taking the cue, I surged back on the trip rope with all my strength. The result was sudden and spectacular. Right on the lip of the dropoff, the mare went straight up in the air, turned a complete forward somersault over the neck yoke, and landed on her back facing us with all four feet in the air. We came to a sudden stop with one horse standing right side up and the other upside down—each one pointing in a different direction.

Dad pushed his hat back, looked at the team and then at me. "When I said whoa," he remarked dryly, "I didn't figure that you would turn the outfit inside out!" But then, looking down over the edge of the coulee, he added, "But not a bad idea in some ways. Not a bad idea at all!"

Then he tipped his hat and grinned at Mother, who had come to the door to survey this latest escapade. She just set her lips in a disapproving line and turned back into the house, shaking her head in resignation. It's a wonder her hair didn't turn prematurely white from watching us work with broncs, for she worried about us, and we sometimes gave her reason to wonder how we managed to stay in one piece.

When we got the mare back on her feet, she stood like a lamb. Apart from a broken belly band and a couple of broken tracehangers, no damage was done. From that day on the mare was very sensitive to the command to stop. All one had to say was something that sounded like whoa and she would come to a skidding halt with her head up and a look in her eye as though she expected the sky to fall on her if she failed to comply on the second.

My father tells of an amusing incident that happened many years ago before the ranges were fenced. Early one morning on spring roundup, the foreman was starting the crew out on "circle"—assigning riders in pairs to various portions of the country to gather all the cattle from that part of the range—when at a distance they chanced to see the cook's four-horse team running away with the chuck wagon. The foreman, along with three or four men on fast horses, immediately rode at top speed to cut off the team and help stop them.

They were hard pressed to catch up, and when they drew close, they were astonished to see the cook sitting up without any reins in his hands, playing the silk popper of a long four-horse whiplash over his team as they burned up the breeze across the prairie with the wagon swaying and bouncing along behind. Every once in a while a wheel would hit a badger hole, the wagon would go extra high, and a bedroll or two would fall off.

Splitting up, the riders converged on the runaway team from both sides and, grabbing bridles and trailing reins, managed to bring the team to a stop.

Angry at such apparently reckless behavior, which could have easily wrecked the very important chuckwagon many miles from repairs or replacement, the foreman eyed the cook in a hostile manner and in a somewhat caustic fashion inquired what he thought he was trying to do.

Cooks being independent and not easily cowed by foremen or anybody else, this one looked his boss right in the eye for a while, and then after a somewhat prolonged and heavy silence that sank in all around, he said, "Well you see, when I started the team off, one of the leaders' reins broke. I figured sure as hell if I hung onto the three that was left, I would do something wrong and pile the outfit up. So I threw the whole works away, and took down the whip to keep them runaway sonsabitches straight!"

Apart from the bedrolls scattered across the prairie for a mile and the slight delay, there were no adverse results of the incident. The cook's reasoning was good, and when somebody laughed the tension of the moment disappeared. The roundup proceeded with the cook enjoying an addition to his reputation. Apart from his already recognized ability with the skillet and Dutch oven, he became well known as a great "skinner" (or teamster).

Some spectacular things could be done by men who knew horses and how to handle them in harness. One day a horse buyer showed up at my grandfather's ranch looking for a matched four-horse team. Dad and his brother Andrew corralled a bunch for him to look over and in due course he picked out four well-matched, four-year-old Clydesdale geldings. The price arrived at was a high one, for a matched team of four is not easy to find. As part of the bargain Dad agreed to drive the horses to town next morning and deliver them directly off the wagon. Nobody mentioned the fact that none of the four geldings was as yet even halter broke.

But this was no problem, for a skillful horseman can halter break as many as six horses a day. As soon as the buyer was out of sight, Dad and his brother started to work on the team and soon had all four leading well. By dark, they had them all tied in the barn, fully harnessed, and with tails, manes, and hooves neatly trimmed.

At daylight next morning they led the horses out into the corral where the wagon waited. By careful maneuvering, they got the team hitched up without any trouble—two and two in line, since four or more horses were always hitched to a wagon. My father climbed up onto the seat of the big double box freight wagon taking a firm grip on the reins with all the horses facing the gate. My uncle eased the gate open and then came back to climb up beside Dad. Picking up the long-lashed four-horse whip, he shook it out straight and then snapped the silk popper like a six-shooter going off just over the horses' backs. In one bound the four were flying out the gate heading for the big river flat. While Dad kept a firm hold on their heads with the reins, Andrew played the whip on them—not punishing them, but just flicking them here and there when necessary to keep them running on course.

They made a circle of about a mile out across the bottoms, and when they peeled off it heading up the road for town the horses were slowing a bit, but there was still a long, lifting plume of dust strung out behind the wagon. By the time they reached town, eleven miles away, the team was going at a slow trot and beginning to answer the bits.

The horse buyer accepted them without question, and they eventually became a dray team hauling a heavy brewery wagon. For years they took ribbons at local fairs, and it is quite likely no judge ever suspected the vigorous short-term beginning they had enjoyed by way of education.

Things did not always proceed so smoothly when it came to breaking and training horses to harness. When I was still just a long-geared kid, I was boarding and going to school from a dairy farm just outside the city of Lethbridge. The farm was owned by a very well-known horseman, Bert Tiffin, who, even now, many years after his death, is remembered for his generous spirit and his remarkable ability with horses.

Bert had a fine-looking black gelding weighing about thirteen hundred pounds, which is too light for heavy work and just right for use on a light delivery or express wagon. The black gelding was partly gentled and trained, but there was little for him to do on the farm, for there was no teammate for him.

But one day Bert heard of another black gelding belonging to a neighbor, a character by the name of Jim Thomson. Jim's weakness was horses, most of whom were of little use to him, for his bunch ran loose on a big piece of open range not far from his small ranch, were of indeterminate breeding and age, and largely wild as deer.

A hired hand and I were sent to round up these horses and corral them at Jim Thomson's place. This we managed to do, but only after some hard riding that took us across miles of prairie, through nests of badger holes and across patches of prickly pear. Had we not been well mounted, I doubt if we could have kept that bunch in sight, for it was led by a pair of thoroughbred mares with sucking colts trailing at their heels—a combination that can make gathering horses a tough job on occasion, since mares with colts are always looking for places to hide. When we finally ran the bunch down the side of Six Mile Coulee onto a little flat where the corral was located, we had them tightly bunched and really flying or we never would have managed to pour them through the gate. As it was, we barely got the gate shut in time to head off some of the bunch that were making a quick break for freedom. Among them was a beautiful coal-black gelding with scarcely a white hair on him.

He was a picture horse, even though he was about as unkempt as a horse can get. He stood a good sixteen hands high, and was deep of barrel with clean, well-muscled legs and a trim head. His mane hung almost to his knees in twisted, tangled ropes of hair all stuck together with cockleburs. His tail almost dragged on the ground, and it was so full of burs it was almost like a club. He stood among the bunch breathing hard, his eyes flashing and his nostrils flaring with nervousness and pride.

Upon Bert's instructions, the hired man went into the corral to rope the black. His first throw was close, but the horse ducked the loop and spun to kick at it in the same motion, moving as though on steel springs and quick as a snake. The second loop also missed, and this time the black tried to jump the fence; but it was too high for him and he fell back among the rest of the bunch on his back. He was up on his feet in a flash, and the way he stood chucking his nose and eying the hired man, it was easy to guess what he had in mind for his next move. The man stood fiddling with his loop, uncertain of what he should do, obviously aware that he was playing with dynamite and not liking any part of it.

Bert called him to the fence and took the rope himself. Then suddenly turning to me, he said, "Here kid, take this and go catch that horse."

Having played and worked with ropes since I was old enough to take notice, I needed no second invitation. Walking toward the bunch at the far end of the corral trailing an open loop in the dust behind me, I watched the black.

He stood facing me with fire showing in his eyes and rollers in his nose, and for a second I thought he was going to rush me; but when I kept coming, one of the mares shouldered into him, and then he whirled away. I was almost within throwing distance and made two or three quick steps just as the black spun on his heels to my right. Another horse got in his way, and he reared trying to get past. His head was sticking up almost stationary as I spread a hollihan flip— a backhand throw Dad had taught me—and dropped the loop over his ears around his neck. It was barely tight when he came down and charged me. Running for the fence, I rolled under it just as he hit the rails over me with a mean grunt. I still had the end of the rope and by the time he recoiled and turned, I had a wrap and a half around the bottom of a stout post.

When he hit the end of the rope, he almost took it away from me and the jerk threw him. I managed to hang on, and before he could get back on his feet, Bert ran in close and jumped on his head to hold him down. We hog-tied him there on the ground.

We trimmed his tail and mane short and curried him clean. Then we put a halter and a pair of hobbles on him and let him up. The trim gave him a different look—now he was more like a civilized horse. Instead of a wild ridge runner, he was now a bit confused and his self-confidence was shaken. He must have felt a bit like Samson after Delilah had worked him over with the shears, for he remembered he had once been trained to halter and followed us home with no trouble.

But he was by no means gentle, and it was several months before Bert let any of the crew drive him and his teammate on the light wagon used for the daily delivery of milk to the dairy about eight miles away. They were a handsome pair of horses hitched up to the wagon in their brass-studded harness—a magnificent pair of animals that trod the earth as though they spurned it and longed to fly. When they passed, people stopped to look and even those who knew nothing of horses remarked on them.

It was not long before the manager of the dairy noticed them and he offered Bert a good price for them. He wanted them for a delivery wagon team in the city, but Bert refused, knowing that the team was still too green for that kind of work. The dairy manager mistook Bert's reason and made a higher offer. When Bert still refused, he was a bit put out and said nothing more about it for a while. But he couldn't keep his eyes off those blacks, and after two or three months had passed he came back with a third offer. It was a fine price for any kind of team and Bert closed the deal, but with the understanding that he could keep the horses and work them till he was sure of them. It was almost a year after he started working this team that he finally delivered them.

In the meantime I was living in the city and going to high school there. Quite by accident I saw the final chapter of this story written in action.

On the way to school one morning, I was approaching an intersection when I heard the unmistakable and familiar drumming of hoofs on the pavement. The next instant the black team came thundering around the corner going flat out with the wagon careening behind them. The driver was nowhere in sight. Fortunately, that early in the morning the traffic was thin in that part of town, for the horses' momentum took them away over to the wrong side of the street as they made the turn. How they managed to hold their footing at the speed they were going is a mystery, but somehow they did. The wagon slewed and skidded in their wake to hit the curb sideways and upset, throwing a shower of glass quart bottles full of milk down the gutter and along the sidewalk. The wagon box came off. The team never slowed up. As a matter of fact, they seemed to pick up speed. The next instant they went around another corner and disappeared, not heading for the big dairy barn, but for open, familiar range. They straddled a big telephone pole on the edge of the city and sheared off what was left of the wagon and most of their harness, but strangely enough neither horse was hurt. When Bert Tiffin located them a couple of days later all they had on was their collars and halters.

That team was never the same again. Bert said they were so spoiled by the runaway that about all a man had to do to start them again was spit. They ended up going to a canning factory.

THE INDIANS were also masters at breaking horses, although their methods differed a little from those of the settlers. Because their horses were their wealth, they kept close watch on them, and the horse herd belonging to a band was almost always under guard. As a result, their horses were constantly aware of man and used to a certain amount of control. Likely they were not as wild, for the most part, as horses belonging to ranches in the early days of white settlement, which were allowed to run free from the time they were foaled till they were caught to be gentled for work. The Indian's approach to his mounts was completely utilitarian, with little or no feeling or affection mixed up in it. To him the horse was cold-blooded currency first—a measure of wealth—something he stole on occasion from other tribes, a sort of prize in a game of sporadic warfare, or a thing of value he could trade to a tribal father for a dusky-skinned maiden to keep his lodge and warm his bed. The Indian's horse was a means to an end. It was an animal that carried him and put a definite mark on his social status. A good war horse or buffalo runner was highly prized and very carefully looked after, but only after proving its value by actual test on the prairie.

When I was a boy I once watched some young Blood men breaking a horse. They had a war bridle on the horse's head with a long lead rope. Mounted double on a gentle horse, they led the wild one out into a big slough till both horses were standing up to their sides in the water. Then one of the young Indians slipped over onto the wild one's back. There is little a bronc can do by way of effective bucking in three feet of mud and water. It just headed for high and dry ground in long plunging jumps. About the time it got firm ground under its feet, the brave on its back slid off, hanging onto the long lead rope. Soon they had led the wild one back into the slough for another round. They kept at it till the bronc was heaving for breath and exhausted. Then they let him travel out on the open prairie, where he went without undue fireworks.

In the old times they sometimes started a new horse by throwing and tying it down. Then a wolf skin was flapped and rubbed all over the animal till it quit struggling. In this way the horse was cured of fear of things flapping around him, and it also learned to ignore frightening smells when in the company of man. It was crude by white man's standards, but like a lot of things, only relative to belief and custom; for while we often tend to be critical and poke fun at other ways of doing things, sometimes we overlook their simplicity and effectiveness. A horse thus treated would tend to look hard before kicking at something strange when being handled by a rider.

As has been previously mentioned, in the few short years after the horse came to the plains and mountains of western

America, the Indians' habits, customs, and social life were revolutionized; the people changed from semi-nomadic foot travelers employing dogs for beasts of burden to proud, free-traveling nomads. They were recognized by military authority of the time as being among the finest cavalry of the world.

The courageous retreat of the Nez Percé tribe under the magnificent leadership of Chief Joseph in 1877 has gone down in the annals of warfare as being one of the classic examples of all time of a well-fought rear-guard action. It is still used today as an example and studied by students in the leading military colleges of the world as one of the great actions of its kind. Not only the men, but the women and children as well, were involved, along with several hundred head of the famous Appaloosa horses that were being driven loose by these people. They fought and outmaneuvered the U.S. cavalry all the way, while traveling through incredibly rough mountains from the Camas Flats on the Snake River in Washington and a thousand miles across some of the most rugged country of the continent, and then traversed another five hundred miles of grasslands in west-central Montana heading for Canada. But bad luck and exhaustion overtook them just a few short miles from the Canadian line, and the Nez Percé were finally defeated.

But in the doing they had written some exciting pages in history—pages of raw courage, intelligence, and tremendous endurance. It was perhaps the greatest field test of a breed of horses, for which the Nez Percé were responsible for developing. The Appaloosas were more than outstanding; they were magnificent in the manner in which they met this great challenge. The left-handed recognition of the military against which they fought almost wiped them out, for following the action, under military recommendation to Congress, a law was passed ordering every Appaloosa horse on the plains to be shot on sight. This shameful, blind scheme almost worked, but thanks to providence and some far-sighted people, the breed survived. As usual, the Nez Percé were mistreated and degraded for their efforts, their land was taken away, and they were relegated to what was thought to be worthless country. But they had the last laugh, for oil was discovered on their portion of desert and they became rich by white man's standards. Furthermore, the Appaloosa horse became synonymous with the Nez Percé name. It is a popular breed among horsemen today, and at this moment some of these spotted horses range my pasture.

4

# Horse Trails and High Country

WHEN I WAS IN MY LATE TEENS, MY FATHER AND I HAD A working arrangement wherein he bought horses here and there when occasion presented a bargain, and I undertook to break and train them to saddle. There was always the distinct possibility when handling such unknown quantities that it would be me who would get "broke." For these horses had spent the first several years of their lives running free on mountain ranges, knowing no feed except what they found, surviving tough winters and knowing little of men. In the

first, second, and even third meeting with such a horse, fireworks was generally expected, and rarely did a bronc-buster
step down out of the saddle after such a ride feeling that he
had been short-changed for action. Some of my father's
bargain horses did about as much to train me as I did for
them, and I developed a keen vision in being able to predict
stormy going by just looking one of them over from across
the corral. The fact that I got half of the profit involved
when we sold them kept me from developing a certain rather
morose outlook on the future prospects of a bronc-buster.
It was a rough way to earn a dollar at best, even if there
was a definite prestige attached to being a rough-string rider,
along with the fact that one had little trouble holding up his
end in storytelling sessions around evening fires. Anyone
who rode raw horses in mountain country never went short
on adventure.

Late one winter Dad bought a big brown gelding from a
neighbor. The horse was just a bit short of being five years
old, stood about sixteen hands high, and weighed eleven
hundred pounds. He was ranging up among some wind-
swept ridges and hills a few miles north of the ranch, and
as the bargain did not include delivery, we had to take
possession of him the best way we could.

It was a clear winter day when we saddled up to go for
him. The mountains looked cold and austere in their mantles
of snow. There were eight inches of snow on the ground,
which creaked under our horses' feet as we rode up through
some benches among winter-bare aspens and willows. Here
and there the surface of the snow was marked by trails of
deer, coyotes, and smaller animals. It was horse tracks we
were looking for, and we had ridden perhaps an hour when
we cut some two or three days old. Horses do not range far
in winter, and it was not very long before we came up to this

bunch pawing and grazing through the snow on a big bench alongside a ridge that had been the location of an early homestead long abandoned. About all that was left of it was the four walls of an old log barn with its roof gone—likely blown away during some wild winter storm years before.

There were about a dozen horses in the bunch—mostly young stock with a couple of old mares running with them. Predominant among them was the big brown gelding. He was a good-looking horse, and even under his long shaggy winter hair it was easy to see he had some quality. The whole bunch had wintered well and had no bones showing through their long coats.

While they stood with their heads up eying us curiously, we sat our saddles on the edge of the meadow wondering if we should drive them down off the meadow to a corral about two miles away, or try to make a catch right there. If we took them to the corral we would have to bring them back after we caught the gelding, for the corral was in another pasture. It would take some time to halter break our new horse as well, which would mean it would be long after dark when we got home, since the days of a northern winter are short. We were both riding grain-fed horses, so we decided to make a try for him there.

With a little luck and some planning we figured we could pull it off. Making use of a couple of loose poles and a pair of old wagon wheels still attached to an axle, we made a crude wing on one side of the old barn door. Tightening our cinches, we mounted to ease around the bunch and began pushing them toward our makeshift corral. Maybe we were lucky, or perhaps this bunch had been inside the old building before—maybe looking for shade to get away from flies the previous summer; anyway an old mare with a yearling colt at her heels walked inside the place as though she owned it, and before the rest of the bunch knew what was happening, we had them all corralled.

The big gelding would have broken past us if he had gotten a chance, but we pushed him in with the rest before he could make up his mind which way to jump. He was nervous and stood with his head up as far away as he could get, his nostrils spread and steam flying with every breath. With his loop ready, Dad blocked the door with his horse while I eased around the outside of the walls with my loop in my hand, also ready for a throw. I planned to rope the bronc and hold him while we let the rest of the bunch go, and then we could halter break him and lead him home. But he seemed to know what I had in mind, for before I could get into position for a good throw, he tried to jump the wall. He hung up on top of it and my loop caught him there, snapping shut around his neck just before he fell outside on his back. At that moment my horse half fell over something hidden in the snow, throwing slack in the rope just as I took my dallies on the saddle horn. Before he could recover, the bronc jumped to his feet and hit the end of the rope with a sudden jar that almost jerked my mount off his feet. The honda (running noose on a lariat) broke, and in one bound the bronc was heading for the horizon with his tail flagged over his back.

We still had the rest of the bunch cornered in the barn, which was an ace, perhaps, for another play. We figured it would not be long before the gelding came back looking for them, for rarely will a young horse of this kind quit mares and colts unless he is driven to it. So while I tied a new honda on the end of my rope, we watched. Sure enough, in about twenty minutes he showed up backtrailing across the flat at a slow walk, sniffing and whinnying for his friends. I kept my mount out of sight on the far side of the building waiting for him, but he was suspicious and would not come in close, so I put my horse into a hard run to try to get a loop on him.

The gelding was fast, and in spite of a great burst of speed my horse put on, I was still about a jump and a half short of getting a throw. We were heading straight for the top edge of a hillside where a meadow sloped steeply down to a belt of heavy aspens in the valley below. I knew that if I was going to get a rope on him it would have to be before he hit the slope, for there a horse carrying weight would be in a bad spot on poor footing and a loose horse would have all the advantage. Right on the edge of the rim I got my chance.

There was a snowdrift piled up by the wind along the edge of the drop-off. The bronc was going flat out and looking back at me when he hit it. He came within a whisker of tail over ears in a fall as he went into the crusted snow clear up to his shoulders. For a second he was hung up, and before he could jump clear, my loop was on him snug just back of his ears.

My horse had no chance to stop, but he was ready for the snowdrift and broke out of it about a stride ahead of the bronc. The bronc tried to tie himself in knots as he felt the rope. He went so high and wild that he rolled clear over, and as he scrambled back onto his feet, my horse hit the rope and jerked him down again.

The ground was frozen like iron under the snow, and the footing on the slope was just as bad as it could get. Knowing my horse would have very little purchase on a lateral pull, I turned him straight down and let him head for low country. Every time the bronc would get his legs under him for a jump away to the side, he got jerked off-balance and would occasionally roll clear over. The snow was cushioning his falls and there was little danger of him getting hurt, so I let my saddle horse use his own methods for playing our catch. After the bronc had gone a couple hundred yards upside down, about as often as he was right side up, he was trailing us with a growing measure of caution. Just on the edge of the timber, I stopped to let him think things over while Dad caught up. Together we hazed him the rest of the way to the bottom, where we broke out of the timber onto an old wagon road.

My blood was up and I was enjoying the excitement—feeling good about pulling off a tough catch and a wild ride. I decided to switch my saddle onto the bronc and ride him the rest of the way home. So we got a hackamore on his head, and Dad snubbed him close to his saddle horn, while I clinched my saddle on his back. Then I stepped up on him and Dad gave me the shank. So many things had happened to that big horse that he was too confused to be able to make up his mind about this added indignity. While my loose horse pointed the way toward home and Dad brought up the rear, we made good time. We arrived back at the home corral without any more hard riding. I had that exhilarating feeling of having come through some fast action without losing any hide, another green horse was well on the way to being gentle, and I was sitting mighty tall in the saddle.

Dad sat his horse grinning to himself as I stepped off the bronc at the barn door. He knew that more than the horse was getting educated, and that there was nothing better to give a youngster the kind of iron needed for living on the edge of the wilds than this type of work. I look back with a feeling of great good luck in being born to parents who understood a boy's love for excitement, who gave me a chance to build and pit my skill against the kind of problems we encountered in a country that could be bland and kind to those who are ready, and most unkind—on occasion even pitilessly ruthless—to those who were not. My brother and I were given a chance to enjoy conquest in a field where laziness or carelessness meant getting skinned up and maybe even some bones broken. We had some close shaves over the years, but neither one of us ever got seriously hurt. We developed and sharpened our wits to a point where we could see danger before it reached us, and it is just this little edge of foresight that kept us out of what sometimes could have been serious trouble.

The big brown was one of the more satisfying encounters along some rugged trails. He turned out to be one of the best horses I ever rode. When he was seven years old, I sold him to an English horse buyer for a good price. He wound up his career as a cavalry mount attached to one of the Imperial regiments in Britain. Doubtless none of his riders ever guessed that he had been foaled in the shadow of the Rockies, for he was answering trumpet calls a long way from the wintry mountain meadow where I first saw and caught him.

WHEN I WAS NINETEEN YEARS OLD I left home to take a job as a rough string rider and packer for a wilderness pack outfit belonging to the recognized dean of outfitters in southwest Alberta, F. H. Riggall. "Bert" Riggall, as he was known to everybody, was something more than just a guide and outfitter; he was a world-recognized botanist and naturalist, author of some note, a great raconteur, and a wonderful teacher. We were from the first something more than employer and employee; we were good friends. He perhaps recognized a kindred spirit where the wild country of the mountains was concerned, for he gave me an education in natural history like nothing that could have been acquired in any other way. Books were but a secondary feature of this association, though important, and the lessons were largely learned from direct contact by example in open country. Under his instruction, I also learned the art of balancing loads on pack horses, thus becoming thoroughly acquainted with several variations and uses of the diamond hitch.

The diamond hitch is a very simple yet complicated manipulation of a forty-five-foot rope attached to a lash cinch at one end, by which various kinds of packs can be tied down to a sawbuck or Decker pack saddle in such a way that a horse or a mule can travel all day carrying a load in rough country without losing it. It is called a diamond hitch because the rope pattern takes the form of a diamond on top of the pack when the slack is pulled out of it. Properly tied on a well-shaped pack that weighs somewhere between 120 and 180 pounds, it is an ingenious and very efficient method of holding a load on the back of a horse or a mule. Carelessly applied or when used with odd-shaped packs, it can be the nemesis of the sunny disposition of a packer and a means of great amusement to pack animals.

The diamond hitch is thought to have been originally used by the Basques in the high mountains of northern Spain, perhaps three thousand years ago. It was brought to America by the Spanish conquistadors and used during the conquests of the great inland empires of the Incas and Aztecs. Later it was adopted by the early free trappers and the gold prospectors of the vast placer strikes of California and other regions of the west. It was used extensively by sheep herders and cattlemen in packing supplies into country too rough for wagons. It was always a major feature of transportation employed by wilderness guides and outfitters of the western mountain regions. No man who has ever used it can claim that he has completely mastered its vagaries, and some have trouble remembering how it is tied in its various forms if they haven't had occasion to use it for over six months. I knew one excellent packer who had to be shown each spring how to tie it. He was a reasonably intelligent man, but over the winter months he somehow lost the thread of how to apply the diamond hitch. To save time, we demonstrated it for him annually, whereupon he immediately proceeded to tie it expertly for the rest of the summer and fall seasons.

A wilderness packtrain may consist of a bunch of fifteen to fifty animals under saddle, some of which are ridden by the people involved and the rest all running free with loads tied down to pack saddles. Properly handled and suitably equipped, such a unit has a minimum of difficulty on the trail, but when one considers some margin of human error mixed with a multiplicity of horse characteristics, the possibilities for interesting sidelights multiply.

A packer never knows when a yellow jacket's nest may get stirred up under the passing hooves, and when this happens results can be sudden and explosive; but looking back over almost thirty years of continuous packing through some of the most rugged country of the world, even this unwelcome contact caused no widespread uncontrollable panic among our horses. As among people, panic can be overcome by a sprinkling of a few cool heads in the crowd. I have seen the calm, firm command of a packer spread a ripple of quiet confidence through an entire string of forty horses on the verge of an equine riot. It was the sound of a familiar voice, and they responded.

Among such a string there is a wide variation of character, which tends to put each animal in its place and keep it there. There are the fast ones and the slow ones, the smart ones and the ones that are less smart. There are the ambitious, courageous ones—the born leaders that the others prefer to follow. There are the drones, and also the mischiefs, which are always looking for ways to stir things up. There are the sensitive ones always quick to take offense.

There was a big bay gelding by the name of Jimmy for example—twelve hundred pounds of beautiful equine dynamite that refused to carry a man but carried a valuable pack without damage to it or its contents, in fact, without damage to anything for eighteen years.

He was one of the rough string during the time I first went to work for Bert Riggall. If I ever laid eyes on a bronc that I longed to ride it was that one. He was a magnificent animal, as proud as Lucifer, and he carried himself as though he had ambitions to join the legendary Pegasus. He had been ridden a little at the age of five and then turned loose. When I first saw him, he had been running loose for a year with nothing to do. Doubtless he considered his brief acquaintance with a rider as something he wanted nothing more to do with, for my first attempt to ride him put me in the midst of the most violent session of bucking I had ever encountered.

He blew up suddenly upon feeling my weight, and his jumps hurt like the blows of a hammer. It was like a double-barrel shotgun going off under the saddle every bump, for he hit hard on stiff legs and popped himself like a whiplash on the top of his leaps. I stayed long enough to find out what real punishment was like, and then he threw me clear to the knot of a twelve-foot buck shank. I was high over his back when I turned back for earth and could see him bucking under me. My landing was mild in comparison to what had been happening in the saddle—so mild it was hardly noticeable.

I tried him again another day, and he threw me hard again. One did not just get up and crawl back on this one, for the riding and the coming down were hard enough to call for some recuperation.

Both of these sessions had been in the open, where all the advantage was with the horse. The third time I got feeling salty enough to fit a ride on him, it was in a big round corral at a cow camp on the Oldman River. This time an interested audience perhaps lent some determination to my efforts.

I do not remember much about that ride except that it was the wildest storm of bucking and swirling clouds of dust ever experienced. It seemed to go on forever. When it finally came to an end, Jimmy stopped with his rump against the fence. I could distinctly feel the beating of his heart between my knees and hear his heavy reaching for wind. There was the sound of my spur rowels jingling softly in time with his heaving and shaking. There was the positive knowledge that if he took as much as one more step I was going to hit the ground. It was a somewhat frightening experience, because it was the first and only time in my life that I was conscious of being totally blind. My eyes registered nothing but a dull gray and I couldn't even see Jimmy's twitching ears.

It was only a few seconds before my vision came back, but they were long moments. When I stepped down off him, I knew he could be ridden, but I also knew he was not for me. From that day till he died he was a pack horse—and a first-class one. He and I made a treaty that hot afternoon in the corral in an aura of the deepest mutual respect, which was kept to the letter for the eighteen years we worked together. Till he died at twenty-four years of age, he still had a proud, unquenchable fire in his eyes, and there was not a blemish on him anywhere. His mighty heart just stopped beating one day, and a great mountain horse lay down for the last time.

There was the unforgettable Amos, who came with me to join the packtrain when he was two years old. Like the "little black bronc with the long shaggy mane" in the cowboy song, Amos was a definite personality among horses. Short-coupled and strong, he was a deceptive horse. He had the wild-eyed look of an incorrigible outlaw, but at heart he was as gentle as a lamb. When he was young he liked to buck, but more out of sheer exuberance than with any desire to unload his rider. He looked small, but he was a strong, well-constructed horse weighing at least eleven hundred pounds. He was a very useful saddle horse and a superlative pack horse, with more than the usual amount of good "horse sense." No matter where or under the most unusual conditions Amos always welcomed the appearance of a person he knew and would come to be caught. This was by long odds the one attribute among his many attractive features that endeared him to my heart, for when a horse was needed in a hurry one could take a bridle and go catch Amos. Anyone who has ever worked with horses in wild country knows how valuable such a horse can be.

Amos had a real zest for life and an enthusiasm about doing his everyday chores that got us in an embarrassing jam one day.

One summer a year or so after I started to work for Bert Riggall, we had a family from Boston with us as guests for a summer sightseeing and fishing trip. They were new to the outfit and proved to be the kind of people we wanted to keep as regular customers. Toward this end Bert gave us all a sort of mountaineer's pep talk one morning shortly after the trip was under way, to the effect that we were required to make the usual good service even better for this party's benefit, and that any little extras we could provide them would be very much appreciated.

One fine morning after we had reached our central camp, where we planned to stay for the duration of the trip, I was saddling Amos to ride out and bring in the rest of the horses. I could see by the look in his eye and the way his near ear drooped in my direction that he was going to warm us both up with some bucking. And when I saw some of the party coming up across the flat for breakfast at the cook tent, I decided to treat them to some spectacular western entertainment.

So I booted the near stirrup and stepped up in the middle of Amos, and he, with a little squeal of pure devilment, bogged his head and soared while I sat up straight and pretty, picture-book style, fitting a classy ride on him. I knew him so well there was no need for me to even watch his head. I looked straight ahead, watching my shadow out of the corner of my eye and noting with satisfaction the open-mouthed expression of admiration on a pretty girl's face as I went by.

I was definitely not watching where we were going, which proved to be a mistake, and Amos couldn't possibly have cared less. Before I realized what was happening he bucked between two trees where the lady of the party had just strung her washing on a line. The clothesline caught on my saddle horn and promptly broke at both ends, whereupon we went out of there with flags of very personal feminine apparel flapping and snapping in the wind on both sides.

Up to then Amos had been only horsing around. Now, completely scared out of his wits, he came apart at the seams in all directions and did things he had never even contemplated before. I had a tough time staying with him, but somehow I managed it. With a great soaring leap he took off the bank and into the creek, where the lady herself was fly fishing for trout. We blew the pool wide open right under her nose, and then dashed on for a quarter of a mile before I got the snorting, wild-eyed Amos stopped.

Considerably chastened, I rode back gathering up various items of trampled apparel. At the stream I found the lady who owned them sitting on a rock all bent over with her face hidden on her knees and her shoulders heaving. I thought she was crying, but when I placed her dirty, bedraggled laundry at her feet with an apology, she looked up through tears of laughter.

"Please don't say another word," she gasped. "I would have come all the way from Boston just to see it!"

Like all pack horses, Amos was a very gregarious animal, loving the company of his own kind and disliking very much to be separated from them for long. Most horses—especially mountain horses, which are particularly accustomed to being a part of a bunch, sometimes go into a complete panic if they find themselves alone and occasionally cause some considerable difficulty for a packtrain.

One fine summer day in southeast British Columbia, Amos found himself all alone. He was very worried. His ears wobbled at half mast and he nickered unhappily. Threading his way at a fast walk through downed logs and big timber, he followed a trail winding down through the Wall Lake basin among some of the roughest and most spectacular country in the Canadian Rockies. The little runnels of sweat that darkened the black coat along his neck and shoulders emphasized his anxiety, for actually the day was cool and the going downhill. His instinct was to gallop, but he went at a walk, his four white stockings flashing. Holding his blazed face low, he sniffed the rail and overhanging brush for scent of his friends. Cautiously, through long habit, he eased the bulky boxes slung on each side of his pack saddle around snags and trees, as though he knew what they contained. Anyone seeing him would have pronounced him lost, but this was not true. The rest of the packtrain was lost and he was looking for it.

Meanwhile, miles ahead, I led the packtrain out of the timber onto the open beach of Cameron Lake on the Alberta side of the Continental Divide. Turning in the saddle, I checked the forty horse lineup behind me, which I was able to see from end to end for the first time since breaking camp that morning. Immediately I missed Amos. A quick pow-wow with my wranglers revealed that no one had seen him since we had taken the trail from our Wall Lake camp. With sinking heart I remembered tying up his halter shank and turning him loose when we had packed him that morning. It had been his turn to be night horse the previous night, and he was hungry. Now I regretted being softhearted about letting him graze, for obviously he had wandered out of ear-shot of the bells and had not heard us leave. There was no telling where he might go in that kind of country, and there was a chance his pack could get him into serious trouble.

Then, too, he was a walking photography store, and the valuable contents of his pack belonged to my guests. The thought of losing the exposed film he carried made me break out in a cold sweat. This was the third week of a twenty-one-day wilderness packtrain trip, and now a successful expedition might come to a gloomy end. Quickly I dispatched a wrangler to find him, while I took the outfit over a range of mountains to a timberline basin campsite perched on an overlook above Boundary Creek Valley.

In the meantime Amos came down out of the basin onto the main trail leading out over Akamina Pass. Here he got the scent of strange horses mixed with those of the trail mates he was following, and even that confusing mixture was fast fading on a warm, dry wind. Coming to a three-way fork, where a shortcut trail headed for Cameron Lake in Waterton Lakes National Park, Amos stopped in momentary indecision. Here he could double back over the summit into Alberta or he could take the shortcut to Cameron Lakes. He chose the main trail over the summit and so missed the wrangler riding hard up our back trail.

With only the musical tinkle of his Swiss bell for company, Amos hurried down off the pass to where the trail ended abruptly on a highway buzzing with Sunday afternoon traffic. Again he was faced with a puzzling choice of directions. To the east and north, outside the far rim of the mountains, the familiar home ranch pasture waited. To the southwest was a vast stretch of peaks beyond the dead end of the highway at Cameron Lake. Most horses would have headed for home or panicked, throwing the pack to the four winds. But not Amos; the wheels of intelligence were turning coolly in his wise head as he turned southwest along the shoulder of the highway, ignoring the cars in his patient search for the missing packtrain.

Earlier in the season I had taken him up this road to retrieve some supplies left at the warden's cabin not far from the lake. Amos remembered the place and turned off the highway, coming to a halt before the cabin door. The warden had just come in off patrol and was preparing something to eat when he heard the bell outside. He opened the door and was greeted by a soft nicker. Amos shook his head as though asking, "Seen a packtrain? I lost one somewhere around here." The warden rubbed his ears and led him back to the corral, where he threw him some hay. Having recognized Amos, the warden saddled up after he had eaten his

# *A Picture Diary*

*He who has not walked alone and fished for trout on a wild river midst peaks beneath a sky adrift with clouds has not really had a look at his beginning or come to fully understand himself.*
*For it is in such unscarred country beyond the marks of wheels that a man really finds himself—knowing the warm feeling in his soul that only fear is the enemy and that true values are not measured in bank accounts cached away in artificial edifices of stone, but in depths of serenity and peace found where air is clean and water flows cold and pure.*

*Grandfather's ranch was close to the Blood Indian Reserve. There as small boys we met the Indians when they came to visit. They did not often smile, these people, for they had known the great freedom and unfettered reaches of the plains before the white man came and the buffalo vanished forever. Even though they now wore the white man's cloth instead of skins, they were still proud people with a certain dignity and a sadness written in the weather wrinkles of their faces. Black Horse and his wife have long since passed to join their ancestors, but sometimes when we look deep into the flames of the evening fire, we see them riding their short-coupled cayuses through the purple shadows of far horizons and we are nostalgic and know a sadness too.*

*Years back when the west was young, the range was wide open, without a barbed wire fence in two thousand miles. Cattle ranged free across the Great Plains region on oceans of grass that had recently been dotted with millions upon millions of buffalo. The cattlemen were kings of industry then, the cowboys on their working crews a breed unto themselves. They were children of their environment, tough, highly skilled, and self-reliant. They lived on horseback, scorning any kind of work that took them afoot.*

The pendulum of prosperity swung according to the demands of Eastern markets and the incidence of hard winters. A cattle baron could be a millionaire in the fall and a pauper in the spring when blizzards blew and prolonged below-zero weather prevailed, for cattle are no way as hardy as buffalo.

The women of that time were brave, wonderfully able to make the best of their surroundings, and practical. They had to be to live. Most of them rode horseback well and nearly all could handle a team. Some of them handled firearms as well as or even better than many men. When my grandmother went out on the range with Grandfather or one of her sons, she enjoyed taking the reins of the well-matched driving team. These were half-bred hackney drivers—a hackney stud crossed with a cayuse mare—great horses for light harness work or saddle.

Those early homesteads along the mountains were a far cry from the convenience enjoyed by the ranch owners today. Home was usually a log cabin, fresh milk came the short-cut way by hand directly from the cow, fresh meat was usually acquired by use of a rifle or shotgun and fuel for heating was obtained by using a sharp ax and a crosscut saw. The nearest doctor was anywhere from ten to a hundred miles away, and unless an injury or sickness was too severe to be handled by backwoods methods, he usually never heard about it. Sometimes when he did find out about an accident or illness, it was too late. Childbirth was usually handled by a local midwife. My wife Kay was born in a tent in a midwife's yard. She and her sister were raised here on the ranch where we still live. Their two brothers died young, one shortly after birth, from pneumonia, and the other at age five, from polio. Both occupy an unmarked grave on the mountain slope overlooking the ranch, a huge boulder their gravestone. Kay's mother, an Irish girl, is pictured milking a cow sometime during the summer of 1906 by the original homestead cabin.

*It was buffalo country and is still cattle country, where the grass waves
in the wind—a land of sudden storms and brilliant sunshine, where
even yet a man can ride a horse out of sight of barbed wire.
It is a living remnant of what has passed, a land of sky and
folded mountains; an immensity where the peaks look
down on animals and men, mere specks moving across the gently
rolling prairie. It is drained by the Waterton River—the
Indians called it the Kootenai—which joins the Belly, the Oldman,
and the St. Mary's to become the South Saskatchewan. It is a place
where the Great Plains come in close to the mountains,
gliding across an ocean of grass without any interference of
foothills, to go swooping up to the rugged eagle aeries of the Rockies.*

*We boys grew up there on horseback, coming to know what a good
day's work meant, learning the power of endurance and the need
to find our own direction without having to ask too often. We came to
understand the virtue of making the best use of the land without damaging
it, the necessity of bending nature without fighting it or breaking it.
Without knowing how or why, we grew up loving the sun and the soil
and the water—the roots of all life.*

*Jimmy was twelve hundred pounds
of high-explosive horseflesh that
refused to be ridden. We argued
about it in the way of horses and men,
and it didn't matter much if I stayed or
got thrown; I lost the argument.
We worked such horses any way we
could in the big mountain country.
Sometimes a rider got hurt, sometimes
he walked a long way back to camp,
and occasionally a horse wound up
crippling himself; but generally a
man and horse learned to get along
if the man was tough and persistent
enough. Jimmy and I compromised and
were good friends for many years.*

*There is the smell of hot irons, burning hair, and sweat at branding time,
when all the new calves are marked with their owner's registered mark.
All livestock purchased by a ranch are covered by a bill of sale and
usually rebranded. It is the way of legally claiming ownership.
Calves are heeled by a roper on horseback, then dragged to the fire,
where a team or teams of wrestlers take over to throw and
hold them while a hot iron brands their hides; bulls are
castrated and every animal is vaccinated. My father usually
handled the irons while my brother and I roped the larger stock by
head and heels. Two good ropers and a brander could go through a lot of cattle*

*in a few hours
this way.
Generally the
ropers changed horses
two or three times in
a long afternoon,
and an extra
set of irons
was used so that one
was always hot. A
good roper could
keep two sets
of wrestlers
busy when branding
calves. If they
hurried, a team
could handle
ten calves in
almost as many
minutes.*

*All my life I have wandered the wild, high places of the Rockies. My trails
have led me from the wilderness of the Salmon River country in
Idaho to the old limestone domes and pinnacles in the north,
where the shores of the Arctic Ocean rim the
continent beneath flaming banners of the aurora
borealis. In all this vastness of mountains, glaciers,
and twisting valleys, the south Canada Rockies
straddling the continental divide between Alberta and
British Columbia are my favorites. Here in places
you can pour a pail of water out on the knife-edge of
the divide to see half of it go tumbling down on the first long jump*

*of its journey to the sea at Hudson's Bay, and the rest of it fall
tumultuously down toward the blue waters of the Pacific. Here
a man can live and travel among the herds of elk, deer, mountain
sheep and goats. He can slip quietly through the willows backing the
shoreline of a lake to watch moose diving for aquatic growth, with only
their ridiculous tails showing above water. He can chuckle at the
comedy of black bear cubs swatting at angry hornets while their
mother tears a nest apart for succulent grubs. Here he
can climb through snow to meet the grizzly in his mountain territory.*

*It has been said
that a horse works
for a man
only in fear of
punishment or
in the hopes of a
reward. This is
not true. If the man
knows just a bit more
than the horse, he can
train him to do most
anything within the
scope of reason. With
some understanding
and patience most
horses can be condi-
tioned to accept
anything asked of
them without
"kicking the lid
off." Here
a horse called
Chief stands quietly
with a raw bearskin,
fresh off the grizzly that
wore it, tied down be-
hind the saddle. In the
wild and rugged,
standing-on-end
country of the Rocky
Mountains such a
one is not only a
good friend,
but one of man's
most valuable
possessions.*

*One of the great things I remember was my first sight of a herd of bighorn rams—up among the peaks against the blue sky—just in back of our wilderness ranch. My brother and I climbed and played in that rocky terrain, and we grew tough and agile. We never could get excited by the idea of killing the wild sheep, although we ate their meat at times and their horns adorn my wall.*
*They walk blithely, almost completely carefree, along airy causeways among the peaks. In summer they feed on exotic alpine plants. They brave the fierce blizzards on the high, wind-swept ridges in winter.*
*They are one good reason for preserving some wilderness country— a wonderfully warm, picturesque, free part of the complicated ecosystem we must strive to preserve, lest we perish with it.*

*Bert Riggall was one of western Canada's great personalities who roamed the Rockies as a professional guide, botanist, naturalist, and explorer. He was my friend and teacher with whom I shared a thousand campfires. In all the years we worked together with the packtrain through the mountain wilderness, working with horses and people, sharing the same tent and frying pan, there was rarely a word of disagreement between us. He taught me how to look and understand something of what I saw. He patiently taught me how to recognize different plants and how to see the interesting character of birds and animals. He encouraged me to interpret my impressions through the art of blending my camera and pen so that other people could come to learn things important about the wild places that fast disappear.*

*In 1903 he left his native England on a trip around the world. He got no farther than the Alberta Rockies, where he spent the rest of his life. He is gone now, but he left a great legacy enjoyed by a host of people. His grandchildren roam the mountains from the high Arctic of Canada to the far ranges of New Zealand, all of them having an inherent love of peaks and mountain clouds.*

*A great part of our lives has been spent under canvas, where our beds were sleeping bags rolled out on evergreen bough mattresses, our stoves made of sheet iron—usually somewhat battered from being banged into trees by carefree packhorses—and the running water, a cold clear stream*

not far from the tent door. It has been a wonderful kind of gypsy life
out in the open wilderness under the big sky—a happy one we
look back on with nostalgia. For we know that happiness is being
healthy, with enough to eat, a warm bed, and a scrap of
canvas to keep out the weather. It is good books to read,
a camera and a pen with which to record interesting things. It is telling
stories around the evening fires, with maybe an owl hooting
somewhere in the background under the purple canopy
of stars. It is watching campfire smoke lift in a twisting
ribbon between sentinel spruces. It is meeting and knowing people
from all corners of the world in such a setting, where all
the frills and gingerbread fall away, revealing men not as gods,
but as truly a part of nature's pattern. Happiness is growing
up in wild country where there are fish to catch, horses to ride,
dogs to hunt with, and butterflies to watch. It is, above all
things, laughter and love. So we grew up and our children were reared,
and maybe with a little luck our grandchildren will enjoy their lives.

They came flashing up from between rocks and from under
overhanging bushes to grab a fly artfully made of fur, feathers,
and steel. Then they fought to escape, bending the little rod, knifing the water,
breaking the reflections into a thousand fragments by their
leaping. Some got away to live another day, but some came to the
angler—enough to fill a big frying pan—but first he spread them out
on a flat rock to admire them even though their marks and colors faded fast.

*The lure of riches from the beaver creeks was what brought the first white men across the plains to the foot of the Rockies. The beaver were thick as flies on a dead buffalo then, but they almost disappeared under the relentless pressure of the trappers and traders. The formation of the national parks and decent conservation practices brought them back. Now the trapper and hunter take them again to keep them in control under a management system that fits the need of the times. Left to themselves, beaver are wasteful creatures, literally eating themselves out of house and home, leaving aspen groves a mess of fallen logs, and choking the creeks with dams to the point of leaving no water for downstream users.*

*A seventy-five-pound beaver makes a heavy, awkward load for a man to pack out of a swamp where it was shot.*

*Those who left their tracks first at the foot of the Shining Mountains were explorers and trappers. We followed their footsteps. In our teen years—when boys today drive high-powered cars, look toward the stars and sometimes destructive revolution for their adventure—we put on moccasins and spent the winters hunting and trapping furbearers. We ate venison jerky on the trail and boiled the noon tea pail over a tiny fire while the trees around us snapped and groaned with the cold. Day after day the rough*

miles unrolled beneath our snowshoes. Home was sometimes a small cabin in a timber grove close to a spring, and sometimes it was the ranch headquarters. There we worked at night fleshing skins and stretching them to dry on specially shaped boards. Once a month or so we shipped our skins away to a distant fur house to be sold. Thus we earned our money, enjoyed the independence of making our own decisions, wild and free as the animals we pursued.

*T*hat day we stalked a great bull elk herding his harem of cows among timberline larches, where the sun filtered down all gold and warm through the yellow leaves of autumn. My hunter friend, fresh from the tension and rush of a great Eastern city, eager for adventure and action, gasped when the bull stepped out from behind a mess of fallen logs to stand framed between two big trees and bugled his challenge to a rival across the valley. At first the hunter just stared and began to tremble, but when he remembered why he was there, steadied and lifted his rifle. At the shot, the bull leapt, spun on his heels, and then went down. The noise brought a big mule

deer buck leaping from his bed in a clump of brush to go in long springy bounds past us down the slope. Again the rifle crashed, echoing against the walls of rock above us, and the buck joined the bull in the happy hunting grounds without even knowing he had taken the step. We skinned out the capes and packed the elk horns and a load of meat on Amos, then headed for camp.

*Twice the great bull elk came down off the mountain and twice my stalk
was foiled by his hearing the creaking of my boots in snow or the snapping
of a tiny twig. The third time he came the same way at night, feeding
across the ranch meadows and bedding down in the morning midst
a vast thicket of willows. This time I took off my boots to go in
stocking feet as quiet as a shadow, feeling every step and taking more than
an hour to go two hundred yards against the wind. Then on the edge of a
small dry slough a tiny twig caught my sleeve and broke. Like magic, a
great set of antlers rose from the willows across the slough. He was
but one step from vanishing when my rifle sights caught his face and
steadied on a small spot of dark-colored nose. At the shot he dropped as though
falling down a well. Coming close, I stood and marveled at the size of him,
and after much work and sweat brought him home for my wife to see and
admire. We had our winter's meat. Kay remarked that the antlers were,
next the soup, the tenderest part of him. He looks down from my wall as I
write this, and there is a kind of mutual bond between us, for we
have seen the mountain—and we were both pretty tough, come to think of it.*

*When I was a professional guide and outfitter ramrodding a forty-
horse packtrain through the Rockies, my crews were made up of
cowboys, trappers, and mountain men. Together we provided a very
specialized kind of service for the cream of the outdoor sporting fraternity
of North America. Levi Ashman was one of the truly remarkable*

characters who worked for me. Another was Dick Schaeffer. One was
a diminutive, five-foot-seven Welshman and the other was a monumental
six-foot-four Canadian cowboy. Levi in his quiet way was a wonderful
storyteller with a keen but puckish sense of humor. He was also fond of
company, even though he trapped alone in the midst of some of British
Columbia's highest mountains. When he got to town, he generally went on a
binge that lasted for days; but even in his cups he was always gentle, not the
slightest bit offensive. Dick also loved a joke, and his roar of
laughter often shook the tent in the evenings when the fire was going
and pipes were lit. He was the most powerful man I ever worked with.
I have seen him take a bronc by the nose with his left hand and put
his right arm around its neck to hold it still by sheer strength while
we put on a pack. Levi has gone up the last trail over the last
divide. Dick rides a tractor, working in construction now, saving his
money to buy a little ranch where he and his wife plan to retire.
Both these men hated to be photographed. But one day when we had
just finished lacing a grizzly skin into a drying frame and cleaning
a moose skull, I captured them on film.

*R*oy Marshall was an old-time cowboy who came up the trail from the south when he was only seven. He helped his older brothers haze a herd of loose horses, while his father and mother drove in a covered wagon. Twenty-five hundred miles they came, from south Texas to the Bow River country near Calgary, Alberta. Roy was one of the finest horsemen and riders I have ever known. He could be very gruff in his manner with men, but always very courtly and mannerly with women. His heart was as big as all outdoors. He was sixty when he worked with me, more than twice my age, a fact that may have bothered him a bit. One day we were working at something in a hunting camp and he suddenly looked at me and snorted, "First time I saw you, you was about big enough to run under a coyote without duckin' yore head. Anybody had told me then I would some day be workin' for you, I would have laughed. Just goes to show you cain't ever tell how far a frog can jump till you set and watch him awhile!"

meal and led him over the pass to my camp on Boundary Creek.

I was never happier to see a horse and a pack, both safe and sound. From what the wrangler and the warden told me, I was able to piece together the story of Amos's adventure and marveled at his cool-headed solving of a situation that could have ended in disaster. To Amos, it was just another challenge to his intelligence, another adventure in a long colorful career of great service.

Although in his younger, more rollicky years Amos enjoyed shaking me up in a bucking match on occasion, he never in his life bucked with a guest or a pack. If anything went wrong with his pack on the trail, he just stopped to wait for someone to come and fix it—a rare pack horse indeed.

We were trailing through some very tough going down the west slope of South Kootenai Pass one day when a portion of the packtrain came to a sudden stop in some tangled alder brush, where the trail was pitched at a steep angle. The wrangler dismounted and circled on foot through the jungle alongside the trail to find out what was holding things up. He found Amos anchored on the trail; a cased guitar that had been tied to the top of his pack had come loose at one end and was hanging down over his heels. Most horses would have kicked it to bits, but Amos chose to stand and wait for help.

Amos and I trailed together through mountain wilderness country for twenty-seven seasons. The last four or five of those years he did nothing but enjoy the country, for his face was gray with age and his step no longer had the spring in it that had marked his enthusiasm for mountain travel for so many summers and falls. Perhaps it would have been better for him to be left at home, but on the only occasion I tried this he set up such a heartbroken fuss at being left behind that I relented. He was just about six months short of being thirty years old when he lay down and quietly died on the trail one warm evening in September. No tombstone marks his grave, but he is still remembered.

Amos's bosom pal for many years on the trail was a big pack horse named Sally, another equine character. Three parts Clydesdale and one part cayuse, she was sixteen hundred pounds of well-directed power and good horse sense. From her long, flowing tail to the fierce-looking mustache on the end of her Roman nose, she was for years a model pack horse. For many seasons she held the responsible job of carrying the valuable cook boxes so necessary for building good meals on the trail.

But for the first few years after I broke her to pack, she was a completely irresponsible and incorrigible clown. Carrying two hundred pounds of canned stuff as though it were so many corn flakes, she would hoist her head and tail and circle the outfit at a dead run. This and other equally carefree antics usually wore off all the labels from the tins in her pack, reducing the cook to tears, for she never knew if she was opening beans, corn, or canned peaches until the lid was off.

Sometimes Sally hit a tree when she was barreling around and the contents of her pack went flying in all directions. I tried various disciplinary measures, which Sally accepted with stoicism but no repentance, as though saddened by my lack of humor. None of them had the desired effect on her harum-scarum ways.

Then I became aware of a habit she had that eventually gave me the key to cracking her crust of irresponsibility. She loved to walk over a little tree, bending it down to scratch her belly. There she would stand with her nose outstretched, her eyes half shut as she swayed back and forth in a perfect ecstasy of enjoyment while the tree scratched an otherwise unreachable portion of her broad belly. So I began regular scratching sessions, which she accepted with huge delight. From then on she was my shadow and my slave—I could do no wrong. When her high spirits prompted her to go romping off with her pack, I would roar at her and she would subside to come back into line like a well-trained dog. She preempted the position on the packtrain just behind my lead horse, which she defended with devotion.

Often when I would step out in front of the tents to look for game or admire the view, I would hear heavy footsteps approaching to the tune of a bronze bell. My view would be suddenly blocked by a great horse, as Sally gently nudged me with a shoulder and sidled into position to get that itchy, inaccessible place on her belly scratched. When I hit the required spot with suitable vigor, she would half close her eyes and twist her mustache in an equine leer of pure satisfaction, and let go a sigh that could be heard for yards around.

SOME HORSES are born timid and uncertain of themselves, others are full of fire and courage, and still others are lazy to the marrow of their bones. Rarely do you find a truly stupid horse. The occasional outlaw is generally one of the smartest, twisted by mishandling, for a man has to know more than a horse to teach him anything and sometimes the man fails to measure up. In training and working with horses I have often been aware of a sympathetic link of understanding—a sort of telepathy between man and beast. While working in the trail with my pack outfit, I often sensed what they intended to do before they did it. And sometimes I have concluded that they knew me better than I knew them.

One fall, to meet an emergency, I had to drive a group of forty loose horses along ninety miles of trail through hill country at the foot of the mountains, and do it alone, starting at sunup. I had Swiss bells of different keys on my leaders—each one of these horses being the leader of a bunch that got on well together, even within the close-knit association of a pack outfit. These horses knew the trail as well or better than I did, but because our route traversed ranching country with numerous wire fences and gates it was sometimes not easy for a lone rider to keep them together.

But by riding my best horses and changing mounts frequently, I made good time. At dark I kept going, and now it was up to the horses, for there is a limit to hard riding at night. We were heading for the home ranch and this undoubtedly helped, for they gave me little trouble. When one of the bells began drifting off the trail one way or another I knew by the tone of it which horse wore it and what to expect. A sharp command usually straightened out the bunch. I gave them plenty of time, just pushing them enough to keep the stragglers from falling too far back.

In the black dark of a couple of hours before midnight, we crossed two rivers and a stretch of timber. At midnight I turned the bunch through a gate into a pasture on my father's ranch seventy-five miles from where I had started the horses out that morning and not one was missing. Next morning at dawn we hit the trail again and arrived home before noon. Such a ride would have been impossible with horses I did not know—or with horses that did not know me.

There have been times, I am sure, when my horses doubted my sanity. Generally, they have put up with my eccentricities, but once they acted on their own initiative to correct a situation they did not like, and we all suffered for it. We were on an October hunting trip in the Flathead country of southeast British Columbia. It was the shank end of the trip when we took advantage of a clear, still day to hunt for goats. When we came back in the evening, I noticed that the horses had hung close to camp all day, and were standing around looking at us as though expecting something.

I was tired and hungry, or otherwise I might have read the signs. Anyway, we tied up a night horse and drove the rest of the bunch out to pasture up on some beaver swamps not far from camp. The next morning we woke to find the tents sagging under a foot of fresh snow and more coming down in a raging blizzard from the northeast. To make matters worse, the only horse we had left was the night horse, for during the night the rest of the bunch had departed for home forty miles to the east over a high trail across the Continental Divide. They had known the storm was coming and had as much as told us. I bitterly regretted not taking the hint, for it took a week to get them back and pack out our outfit. There was well over 150 miles of riding involved, through country where the snow was drifted seven feet deep in places and thirty inches deep almost everywhere else. It was a grim and exhausting windup to a long season.

ONLY ONE HORSE in hundreds combines the qualities of a good leader on a wilderness pack outfit. I have been particularly fortunate, for I have owned two outstanding leaders of great character and staying power.

There was Elk, a big, handsome gray with lemon-colored freckles. Tall and powerful and weighing about thirteen hundred pounds, he combined cool-headed courage with great stamina. He carried me across thousands of miles of mountain trails, and never once in all the years I rode him did he question my decisions. Almost always I left the choice of the trail to him, but occasionally in some tricky spot it was necessary for me to choose in deference to the capabilities of some of the lesser horses in the string coming behind. His instant cooperation was always faultless. This is extremely important, for a show of fear—or even momentary indecision—will telegraph itself the whole length of a packtrain, causing confusion and sometimes even danger.

Elk's cool courage when the chips were down was something to marvel at, and I will never forget one spot where his willingness to face danger got us out of what could have been a really bad fix. I was contracted to guide a party of geologists through a stretch of trackless wilderness on the upper reaches of the Flathead River. It was the roughest kind of country, complicated by vast stretches of blowndown fire-killed timber, desperately difficult for horses.

One fine, hot August morning I was threading my way ahead of two geologists toward a peak at the far end of a twisted mountain canyon. It was so hot and dry that the last thing I expected to see out in the open was a bear. But suddenly there appeared on a rocky rim a quarter mile above us the familiar outline of a grizzly. Generally a grizzly will make himself scarce at sight of riders, but this one immediately started toward us at a slow lope. Knowing how shortsighted and curious grizzlies are, I was unconcerned at first, thinking he was just coming for a closer look. But the closer he came, the faster he moved, and it occurred to me that he intended to drive us out. With two green riders behind me, who were real sailors on horseback, there was going to be trouble if their mounts stampeded and began jumping deadfall timber.

There are times when a bold attack is the best defense, even when the odds seem heavy against it. So I pulled a heavy Colt six-shooter out of its holster on my belt, reined Elk around, and spurred him straight at the charging grizzly with a great war whoop. Without the slightest hesitation the horse plunged swiftly over the logs, closing the range swiftly. About half a jump away from being sure my bluff had been called, I saw the grizzly suddenly skid to a stop about fifty feet away.

He was well above us on a rock ledge, looking ferocious as he swung back and forth with his head hanging low, champing his jaws and growling.

Elk stood facing the bear as motionless as a marble statue, while I told the big bruiser what I thought of his intolerable manners, his ancestry, and his general character deficiencies. The pistol felt and looked about as potent as a peashooter in this kind of company, and I fervently hoped I would not have to use it. If Elk had so much as flinched anything could have happened. But he stood like a rock, and finally the grizzly began to cool off. After several interminable minutes his back hair began to settle down and he moved back a few feet to the top of a boulder, where he lay like a big dog on his belly looking us over. Finally he headed back up the mountain at a slow walk and disappeared.

With the grizzly gone, Elk, who had given no sign of worry, let out a gigantic sigh that creaked the saddle under me. Such courage and acceptance of a rider's judgment are rare and wonderful.

During a thunderstorm some years ago Elk took shelter under a big spruce on a steep mountain meadow. When lightning struck the tree, he was killed instantly. The thunder, rolling and booming among the peaks, played a fitting requiem for a great mountain horse.

His successor, Ace, made a fair bid to equal him. Ace was a powerful black, as active as a cat and the best climber I have ever ridden. But his special talent was his memory for trails. His qualities of leadership were accented by a natural arrogance, which resented any interference from me in the choice of ground, and I always had to be careful how I bent him to my wishes. Too pointed an insistence on my part sparked his temper. But once over a trail he never forgot it, and I have seen him find and hold a trail under the most trying conditions.

One September we finished a hunting trip on a high plateau a few miles west of the British Columbia border. I had another party coming in immediately, so I left the stoves and tents set up to save extra packing on the return trip. But before we could get back, a short, fierce blizzard blew in from the north, slowing our schedule for the return trip.

It was dusk and we were still miles to the east of the divide, with a rugged eight thousand-foot pass between us and camp. Breaking trail with Ace, I had to choose between making camp in the snow without tents and going over the pass in the dark. If there had been plenty of horse feed, I would have camped, toughing out the night around an open fire, but there was none. My guests, Warren Page, the well-known gun editor, and his two friends, were game to make the try for camp, so we kept going.

But as we swung the outfit up the long switchbacks, bucking deeper and deeper snow, we ran into the heavier gloom of fog, and I began to regret my decision. But at timberline we suddenly ran out of the dark mist into brilliant moonlight, which was encouraging except that it revealed a chilling sight. Here the trail climbed up along a steep mountain ridge crest to weave across a ledgy face to a shallow saddle; then it climbed up past the ragged edge of a steep-pitched boulder field to the summit. A single slip could mean disaster; it was an ominous picture. Up there the wind had drifted the snow completely, hiding the trail and sculpting the scrubby trees in bizarre forms of ghostly white.

Ace stopped to catch his wind and I sat sizing up the mountain ahead. Then the big black began working out the trail. How he did it is still a mystery to me, but he casually ambled up that first pitch over the shelves through the snow-choked scrub without one hesitant step. At the rim of the saddle he rammed through a shoulder-deep drift and halted on the ridge comb. When I looked back at the long string of horses coming behind me I could have cheered, for every one was moving steadily, without a sign of concern.

But the worst was yet to come—where the trail staggered up along the broken fangs of the drift-choked boulder field. Again Ace moved out. With head held low he eased into his work carefully, while I rode with a loose rein, giving him complete freedom. Where the trail lifted around a point of leg-breaking boulders, he suddenly checked and pulled back, then reached out to paw the snow with a hoof. For a moment I thought he had lost the trail, but when I got down to feel around with my feet I found a flat slab of rock had somehow slipped across the trail before the storm and lay across it at a steep angle. Hidden by the snow, it would have thrown any horse stepping on it. But somehow Ace had detected it in time. I slid the rock down the slope and got back up in the saddle. From there to the summit the trail was easy, and when I reined in Ace at the top, the sky was a faultless canopy of stars cleaved by snow-shrouded peaks lifting in great ranks through soft masses of moonlit valley mist. Faint and far away a coyote mourned in a high, lonesome solo.

Ace broke the spell with a gusty snort and a toss of his proud head. Impatient to get to camp, he started down the slope to the accompaniment of the packtrain bells' soft music. If a man is honest, he knows what he owes his horse in a place like this even if he gets the lion's share of the credit for a superlative job of guiding. Reaching forward I stroked his proud arched neck—a salute to a courageous heart.

# The Remittance Men

LOOKING BACK ACROSS HALF A CENTURY TO THOSE FIRST YEARS makes one aware that education can take many forms for a youngster, and sometimes the best of it is not learned while seated at a school desk, especially on the frontier. Much of its accumulation was so subtle and so completely disguised as something else that we weren't even aware of it; and it's likely that this made it even more appreciated and useful when we came to put it to work for us. Certainly fate and luck played an important part.

A few months ago I had the good fortune to share some talk with a remarkable old mountain man who had seen more than ninety winters. He had come from England as a young man by boat to the west coast by way of the Panama Canal. There a political uproar of some kind stopped the boat, whereupon a deadly kind of cholera began killing passengers and crew like flies. But my friend and two others missed it, for they had bought a canoe and gone up a river into the jungle to explore. When they returned their boat was gone, but they signed on with a sealing ship and sailed north. Eventually they arrived in the strait just outside Vancouver harbor in the dead of night and were waiting for daylight before sailing in to anchor. There a big ship almost ran them down, and they were saved by a hair when an Indian in the crew grabbed some oily cotton waste and lit it. The steamer's lookout saw the flames, and the big boat veered away at the last minute to avoid cutting their small ship in two. Then my friend went ashore to celebrate, and he made such a good job of it that the ship sailed without him. Again his luck and a sheer whim of fate saved his skin, for his boat got too close to forbidden Russian waters and none of the crew was ever heard of again. Three or four years later, while packing a bunch of horses in the interior, he met a man who offered him a part in a high-paying enterprise he was planning for the following spring. One thing and another prevented my friend from taking advantage of this offer, and one day he heard how his acquaintance had been jailed following a train holdup.

"It's a sure bet," he assured me, "a man's luck has to be good. If mine had been anything else I would have been dead long ago!"

It was not hard for me to believe him, for my luck has been good too.

PERHAPS THE MOST important thing we learned in those sometimes not too well-appointed one-room schools was to read. And this puts one on the trail of good books, where one book leads to another, and where one's sheer enjoyment fans the spark of ideas into a glow. So a sense of the power of expression through words is awakened, the shining art of communication.

Strangely enough, the old cow camps on the prairies had a fair sprinkling of well-educated men scattered through them—men who had migrated to this big land from all corners of the world. Some of the mountain men also; like Joe Cosley, were products of colleges. I knew one who was a qualified medical doctor, a man who had come west for a holiday after his graduation and had never returned. The spell of the mountains had infected his blood, and when I met him fifty years later he was a happy prospector in the heart of the Yukon mountains. I brought him a sack of mail and it included several scientific journals and magazines. He died not long ago on the Queen Charlotte Islands off the British Columbia coast, a man well over eighty.

As boys we met many kinds of people, from the totally illiterate to the highly educated. Most of the latter were avid readers and many of them played some musical instrument with talent.

Two such men were good friends of ours, and even if they did not realize it they made a heavy contribution to our education. We talked with them, read their books, learned to appreciate good music, shared their hunting fires, and enjoyed their fine hospitality many times. Both were English remittance men. One was Harold Butcher and the other Lionel Brook.

Both men were somewhat eccentric, true characters, with a genius for being original in their way of doing things. They had a ready wit and to be with them was a constant invitation for experience and adventure. A big book could be written about either one, and it would certainly not lack for color and action.

THE REMITTANCE MEN were laughed at, scorned, and largely overlooked by history, yet they made their contribution to the development of the frontier. They came from the aristocracy of Britain, the discards of an old time-honored blue-blood tradition, wherein the lands, fortune, and titles fell as inheritance to the oldest son regardless of his qualifications, and the younger brothers were sent to the colonies—sometimes with a quarterly stipend to help guarantee that they stayed there. There were also "black sheep"—the independent ones whose ways of revolting against English traditions were judged completely unacceptable to the rigid regulations of their society, and so, whether they were oldest sons or not, these men were invited to leave permanently for some portion of the royal realm far away.

They were called remittance men because it was also the practice of British aristocracy to settle a quarterly stipend on their banished sons, which often was the completing touch to their utter ruin, and sometimes the means of their salvation and fortune. Attracted by the promise of adventure and plenty of room to kick up their uninhibited heels, many of them came to Canada. Almost to the man they tried their hands at ranching, although some became law enforcement officers or joined the militia and others drifted into obscurity with no attempt to maintain a job or develop a business. But even these nonentities made their contributions to the development of a big new land.

They loved excitement and had the money and leisure time to pursue it. For the most part, the frontier had about everything required by the early settlers except cash, and the English remittance men helped immensely to fill this gap by spending their money with zest and an open hand. When one chose to go ranching nothing but the best would do. First-class breeding stock was often imported from Britain, and this helped bring up the standards of livestock in the western plains country. In everything they did, from ranching to having a good time, these men spent money as though they had invented the whole idea; and the fact that they put considerable sums of badly needed cash into circulation probably advanced many communities by twenty years in their development.

While many of these inexperienced, somewhat "green" young men made rather pitiful exhibitions of themselves, trying to win the favor and recognition of a people who for the most part held them in ridicule, most of the cowboys and pioneers aided and abetted them in the promotion of the squandering of fortunes.

One man was sent out from England to try his hand at ranching and spent an estimated hundred thousand dollars— largely from a headquarters established in the bar of a hotel. From this place of good cheer in somewhat "dry" periods between checks from his father, he wrote highly literary and glowing accounts of the development of a great ranch stocked with fine cattle and blooded horses in this land of promise.

But one day the news came that his father was coming to

spend some time at the ranch, and he was slightly perturbed at this unexpected turn of events until he thought of an ingenious scheme of renting a ranch, including horses, stock, and everything else, for the duration of his father's visit, even to the point of paying the real owner handsomely to act as "foreman." In due course the father arrived, and the whole scheme worked beautifully, for this staid and conservative old British businessman looked and observed, and was fooled completely. Everyone in the neighborhood was aware of the caper and considered it a great joke. To the man, the neighbors fell in to help with the deception by treating the son as a man of weight and respect. Finally the old man left, but not before tipping the owner of the ranch handsomely and bestowing a check for ten thousand dollars to his promising offspring's account at the local bank, which that inestimable gentleman proceeded to spend as fast as possible on riotous living just as soon as his father was well out of sight.

This story has a somewhat unusual twist and ending, however, for after having spent his money, and becoming little more than a beggar, this man took sick and went into the hospital. There he met and fell in love with a pretty nurse, whom he later married. Fortunately for his new wife, he had a loving and very wealthy aunt, who suddenly passed away leaving him a fortune. This drew the spendthrift back to England, where he remained to become a highly respected and successful pillar of the high-class community.

Another, less cheerful, story is told of a married couple who came to Canada with a generous stipend from offended relatives, who wished to see the last of them for good. They had disgraced themselves by heavy drinking, and upon arrival in Alberta, they proceeded to really blow the lid off, long since leaving any inhibitions they might have had far behind. First they bought a ranch, which they improved with a sumptuously furnished house. Then they purchased a thousand-dollar bull and a considerable number of pure-bred hogs as basic breeding stock for the livestock enterprise. Of course they also stocked their cellar with the finest spirits.

As a sort of celebration, they put on a great party for some friends chosen for their similar appetites for strong drink. Sometime during the night some brilliant character thought up the idea of bringing the bull into the house and making it sing to the accompaniment of the piano. No sooner was the idea expressed than it was done. The bull was led from his snug stall, prodded up the steps onto the veranda, and pushed and shoved bodily through the front door into the parlor, where the lady of the house, an accomplished musician, waited. The huge beast was tied to the leg of the instrument, and the ringing chords had scarcely commenced when he spotted an open window. Recognizing an avenue of escape, he surged toward it dragging the grand piano and completely ruined it as he jumped through the sash. He also ruined his neck, for the fall broke it.

For a while the party was a bit subdued, but a few rounds of the jug soon corrected this deplorable state of affairs, and then the thinker of the group was assailed with another grand idea. They were all getting hungry, so why not have some fried pig's liver? By the light of some dim coal-oil lanterns, the entire party moved out to the pig pens, where in an orgy of slaughter they killed the valuable porkers and removed their livers. The party lasted for days, while the bodies of the bull and the pigs rotted.

In due course of some years of dissipation, the lady of the house died, and her husband went from bad to worse. Their only son became an early drunkard, and finally about the only thing left of the family was a somewhat unsavory memory.

THE COWBOYS WITH which the young remittance men chose to fraternize very often played jokes on them, which were well spiced with the rough-cut humor of the frontier.

Once a party of holidaying cowboys was drinking in a hotel bar in the company of a newly arrived remittance man. A very pleasant evening was enjoyed by all at the expense of the Englishman. But at closing time they allowed themselves to be pushed out the door onto the street, and the door locked behind them, before they suddenly realized they had failed to stock up with enough bottled joy juice to keep away the chill for the remainder of the night. One of them suddenly remembered he had a flask of whisky in his room, number 40, adjacent to the bridal suite where a honeymoon couple had just taken up quarters. The groom was a large, thick, very muscular, somewhat florid young man, who, in the face of the briefest glance, gave the impression of being not only powerful but also very impulsive.

So when the remittance man drawled good-naturedly, "I'll go and get your bottle," temptation reared its head, for the cowboy remembered seeing the honeymoon couple going to their room an hour or so before.

"All right," he said. "It's number 41."

What occurred next is a bit hazy in exact detail, but as near as history's somewhat disjointed record shows, the well-meaning messenger walked right into the room where the groom was taking a "nightcap" from a whisky bottle just previous to retiring. As is usually the case when a bride-groom sees a total stranger boldly enter his bridal chambers without knocking, he is not pleased, and this particular one showed no inclination to be any different. Holding his bottle loosely in his hand, he stood glowering down at the intruder. Without any preamble, the messenger walked right up and plucked the bottle from his fingers with a short "Beg pardon—"

Down at the foot of the stairs, the listeners heard some gruff and mixed-up conversation punctuated by the scraping of boots and noises of furniture being knocked about. Then a door burst open, and the body of the messenger came hurtling down the stairs to land among them.

Picking himself up rather painfully from the hard floor, the messenger ruefully eyed the jagged neck of the bottle still clasped in his hand, and said apologetically, "There is a bloody bounder in your room and he insists on keeping your whisky!"

From the darkness above came a rumble of sheer menace, "Bounder, be I? Wait jest a minute and we'll see who does the boundin'!"

There came the clumping of great feet and the creak of stairs, whereupon the messenger suddenly departed in the direction of the open prairie.

MY FRIEND BUTCHER was the son of a British shipowner who had quarreled bitterly with his father over his desire to go to sea and become a merchant marine captain. As sometimes happens, the disagreement went past the point of no return, and Butcher left the home fires to come to Canada, where he eventually came to own a sizable ranch just above our place on Drywood Creek. He was a fairly successful cattleman, albeit a bit eccentric one on occasion. He was very impulsive and given to either effervescing in good spirits or wallowing in despair and self-pity. He had a sharp wit and a confessed antipathy toward women, which was really only a front to cover his shyness. He had a tremendous appreciation for good literature and music, and in the days preceding radio he had the finest Gramophone money could buy and a collection of classical records numbering in the hundreds. He also owned a very fine library, including the complete works of such authors as Thackeray, Kipling, and Dickens in fine calfskin-bound editions. Among the furnishings of his unique ranch house was the only player piano of the country along with a considerable collection of music rolls for it. On occasion, although he could not read a note of music, he would sit down at this instrument and improvise by ear. At the Christmas concerts held at the local schoolhouse—a traditional entertainment enjoyed by all for more than a generation—he could be prevailed upon to play for the dance that inevitably followed the stage plays and recitations presented by the schoolchildren. With the help of a neighbor who played the fiddle and another who played a guitar, he would tease some lively music out of the ancient organ that would set the feet of the dancers going.

Butcher was a man of many parts, with a crusty exterior that camouflaged a heart of gold. His generosity was sometimes almost an embarrassment to his neighbors. If he entered into something of interest to him, it was in a wholehearted fashion. There was no middle of the road for Butcher even if he did proclaim to be of conservative nature.

Once or twice a year Butcher invited all his friends from near and far to a grand whist party held at his ranch house. This was something more than just a card party, for it provided him with a way to hold up his end in the social scheme of things. Because he was a dry-humored, entertaining man with a great deal of charm when he chose to exercise it, he was a popular guest at the homes of the various ranchers in the country. And in many ways he was a brilliant host, who could mix a piano concerto, "improvised on the instant," with a positive genius for telling an amusing story. No one who ever attended one of these affairs went home with the feeling of having wasted an evening. He usually held this party in the fall.

Shortly before Christmas it was his habit to take a team and sleigh to town. When I was a boy, he sometimes hired me to drive for him and accompany him everywhere, as he went up one side of the street and down the other, greeting all the storekeepers and neighbors encountered with cheerful compliments of the season and paying all his outstanding bills. On one particular occasion he met a lady—a friend's wife, who trilled happily at the sight of him.

"Oh, I'm so glad to see you, Mr. Butcher," she said with a bright smile. "We so much enjoyed your party last fall, and I meant to write you and thank you, but somehow never got around to it. I hope you will accept a turkey for a Christmas present."

Butcher graciously accepted and we all proceeded to the livery stable to transfer the turkey from their sleigh to ours. It proved to be a large one, all gaily wrapped with colored paper and ribbons. Butcher was in a particularly fine humor as the team trotted briskly for home under the bright winter sun.

Every Christmas he put on a noonday dinner party for all his bachelor friends preceded by a gala Christmas Eve party, which was conducted while Butcher prepared for the big feed. His Christmas stag party menu was more or less fixed, for while Butcher was a fair kind of cook, he was not a particularly versatile one. It generally included roast turkey, cranberry sauce, mashed potatoes, canned tomatoes, lots of gravy, and finally, large portions of plum pudding prepared weeks before and kept frozen. There were very few variations.

On this occasion the gift turkey was popped into the oven early Christmas morning and by noon it was done to a rich golden turn. It held the place of honor on a huge platter at the head of the table. Lionel Brook, a perennial guest at this particular party, upon being requested to do the honors of carving, took his place and began to wield the carving knife. Slabs of well-done meat fell away from the keen edge of the blade, as plate after plate was passed to Butcher at the opposite end of the board to receive its portions of trimmings.

Finally only Butcher was left. When he received his plate he looked at it with a critical eye and said shortly, "Brook, I want some stuffing if you please."

So the plate went back to Brook, who looked long and hard at Butcher and said, "Very well, Butch, old top, you shall have some stuffing!"

Whereupon he took the fork and gouged into the carcass of the turkey to come forth with a great load of gizzard and original plumbing natural to the bird, and set it in the middle of his host's plate.

For a few seconds there was silence as everyone at the table gazed at Butcher's "stuffing" with disbelief. Then came a great roar of laughter. They were a rough crew with hearts of gold, and were not about to let such a small thing upset their appetites. Besides, there is nothing like a few tots of Christmas cheer to sharpen one's appreciation of a good joke.

Without a word Butcher quietly removed the plate, got another, and loaded it. As he fell to with his guests to eat heartily, his only comment was, "Nobody but a woman would give you a Christmas turkey that had not been dressed!"

Actually, long habit and the previous evening's liquid refreshments had caught up to him, for ordinarily he bought a turkey at the butcher shop all drawn and ready to stuff. Because he had been unwrapping an oven-ready turkey for years, he had assumed this was already prepared for cooking. And no doubt he had not looked very closely through the alcoholic haze.

Butcher could be very impetuous at times, which was inclined to get him into trouble. One year, to create a sort of conversation piece, he instructed his haying crew to put all the hay in one gigantic stack close to the buildings. Some of the neighbors criticized the idea because of the possibility of fire burning up his stack and his buildings all at one time, and so setting him out in the cold as well as putting his cattle out of feed. No such disaster occurred, but something did happen that wasn't any good.

A strong pole fence enclosed the stack with a big pole gate at one end of the yard to allow access with teams and hay racks. When cold weather came and the cattle were being fed close to the buildings, an enterprising Shorthorn bull figured out how to get into the stack yard. He simply put his head under the gate, lifted it off its hinges, and let it fall where it would. Next morning Butcher looked out to see all his cattle in the stack yard.

Although Butcher set about reconstructing and hanging the gate, the bull never noticed the difference, and the next morning the yard was again full of cattle. After about four such mornings in a row, Butcher's temper was wearing thin. He was all alone at the time, so he had nobody to help or advise him. Swearing to stop the bull or kill him, he took his double-barreled shotgun, tied it to two stakes a few yards inside the yard with its barrels pointing toward the gate. Then he ran a string from the gate to the trigger, loaded the gun, and cocked it. Leaving this lethal trap set, he went to bed.

For some reason the bull did not go near the yard that night. About midnight Frenchy Rivière rode in on his way home from town leading a pack horse. It was snowing and blowing snow on a cold wind with a pale moon shining through a skiff of clouds. It was a thoroughly miserable night and after several hours in the saddle Frenchy had decided to hole up with Butcher till morning. But instead of putting his horses in the barn, as anyone else but Frenchy would have done, he proceeded to turn them into the stack yard with over a hundred ton of hay instead of a mere manger full. But when he went to open the gate, he was greeted by the blast of a gun, orange flames stabbed at him out of the gloom, and a double charge of birdshot nearly blew his leg off. It narrowly missed him and went between his horses. For a while Frenchy stood very still, no doubt dazed. But then he saw the moonlight glint on the barrels of the shotgun and he also spotted the string. Upon circling cautiously and investigating, he found Butcher's set-gun. By this time Frenchy was in a towering rage, and there was a great deal of him to get angry. He stood six feet four in his moccasins and weighed about 225 pounds. Alongside Butcher's five feet nine and 145 pounds, he looked like a giant.

The first inkling Butcher had that he had a visitor was when he found himself plucked bodily from bed by the front of his nightshirt and suspended a good six inches from the floor. In a soft voice that was dreadful in its menace, Frenchy told him what he thought of people who tried to kill their neighbors, while he gently waved Butcher back and forth as though he was of a mind to throw him bodily out the window. Frenchy barely managed to restrain an impulse to take Butcher apart, but he was most emphatic in his personal description of fools who were careless with guns, and no lame explanations or excuses deterred him.

It so happened that I rode up to Butcher's ranch the next morning with a message for him. After knocking twice with no effect, I finally managed to get an invitation to enter. I found Butcher in front of the fire in his living room—something he almost never did before his morning chores were done. Obviously he was upset about something. He did not greet me or turn his head, but just sat looking straight ahead at a spot on the wall. Knowing him well enough to understand how to treat the situation, I sat down back in a corner to read a book for a while, waiting for him to start talking when he was ready. A big clock on top of the piano was all that broke the stillness for the better part of half an hour.

Then Butcher suddenly exclaimed, "Frenchy is an unreasonable man! He is a frightfully unreasonable man!"

Then he began unfolding the story in rapid, excited bursts, but as he progressed with it, the temptation to elaborate took hold of him. No point in letting a good story go dry for want of some telling, so he dressed it up and set it forth in style, and by the time he was done with it, the funny aspects of the whole adventure took hold of him, his usual appreciation for the ridiculous towered, and he laughed till the tears ran down his face.

But then the memory of the indignity of being plucked bodily from bed must have come back, for Butcher suddenly sobered to reiterate most emphatically that Frenchy was a "frightfully unreasonable man."

Frenchy may well have been just that on occasion, even

if this was not one of them, but there were definitely times when he had lots of company. I restrained the impulse to remark on this, however, and having delivered the message, went home with a new story to tell.

There have been those who have accused us people of the mountain country of being prone to exaggerate the stories we tell around the evening fires. They have neither appreciation or understanding, for if they did they would know that telling a story is like selling a horse—the appreciation of a good animal is likely to be much higher if one takes the time and trouble to do a bit of currying and brushing. It doesn't matter how good the basic article may be if it is displayed in mediocrity, a fact of life that all good horse-traders and storytellers are aware of. In the days when the horse still reigned as king of transportation along the foot of the Alberta Rockies, we entertained ourselves thus. For then the radio was but an infant idea contained in a black box full of unexpected squawks accompanied by voices and music—a very expensive box that only a few could afford. There was no television—it had not been thought of as yet. We saw a motion picture perhaps once a year on a rare visit to the city. So when we got together, we sang and told stories around the evening fires. If a man was inclined to hold his end up in local entertainment, he told his stories well and it was an art enjoyed by many. We had no reason to fabricate unduly, for we lived the things we talked about; and when things threatened to get dull or when repetition showed signs of taking the zest from the repertoire, somebody always managed to come up with something new by way of excitement. Butcher, by accident or deliberate design, was always a source of hilarious material.

His reputation for being somewhat precipitate and head-long in his use of firearms had precedent stemming from an episode years before the one concerning the bull and the set-gun. Like most young Englishmen, when he first arrived from the "old country" he acquired several pistols of various sizes and calibers. When he took up his homestead, he went around for some time armed with a revolver strapped to his waist. Now there are few things more frustrating to a would-be gunman than to find nothing on which to unlimber his guns. He may not be fully aware of this condition, but it creeps up on him subtly, and sooner or later he finds an excuse to bring forth the artillery and consequently succeeds in scaring the wits out of someone or damaging something.

Butcher's experience followed this pattern. One fine summer day he rode over to a neighbor's homestead for a visit. As usual, he was wearing his revolver. He had lunch with his friend, a man by the name of Thornley, also recently from England. Following their meal, they stretched out on the big double bed for a smoke while comparing the merits of their sidearms. Some blue bottle flies were buzzing among the rafters of the cabin roof, and Butcher was suddenly seized by an impulse. Carefully sighting on one, he fired.

His friend, somewhat astonished at this turn of events, nevertheless rose to the occasion with grace and a strong sporting instinct. He too raised his pistol and shot at a fly. So a historic shooting match began.

At the height of it, a rancher who lived several miles away happened to be riding in the hills looking for some stock when he heard the unmistakable sound of sporadic gunfire in the distance. The fusillade continued, and his curiosity prompted him to investigate. Emerging from the timber onto the open flat where Thornley's cabin stood, he saw smoke coming out of the open door and heard pistols going off inside. Following each shot there was a howl of derision or triumph. It sounded as though a prolonged gun battle was being fought by two indestructible desperadoes. The rancher left his horse tied in the timber and crawled closer on hands and knees till he was finally in a position laying flat on his belly to peek over the doorstep into the cabin. Dimly through the smoke he spied the two "combatants" lying side by side shooting holes in the cabin roof— a roof that by this time looked like a sieve. So another story about "crazy Englishmen" was born—with some justification, it might be well to add.

When I heard Butcher tell his version of this story many years later, I asked what Thornley did about his roof next time it rained.

"Why the blighter came over and stayed with me," Butcher recalled. "Rotten bit of luck that," he remarked further. "You see, the careless bounder had got himself lousy somewhere in the meantime, and after he went home again, I discovered he had left some of his livestock behind. Horrible little devils—lice—do remarkably well on new pasture —increase at a fantastic rate! Took some doing to get rid of them, I can tell you! Never did quite trust Thornley again!"

While all the Englishmen that homesteaded in the foothill country along the edge of the Rockies were not remittance men, they all shared some original ways of doing things that sometimes nonplussed the "natives."

One such homesteader built his cabin along a trail that led across the narrow neck of a slough, so that when he rode or drove west from his door, it was necessary for his saddle horse or team to fight through this bog. This was especially bad for the good clothes he wore, when he went on Sundays to visit a rancher's daughter. So one day this man industriously undertook to build a floating bridge across this bad place in the trail. The idea was excellent, but he laid the stringers crosswise and spiked down a pole deck lengthwise on these. Poplar poles are at best not very straight, so there were gaps of various lengths and widths running across the bridge. To make the somewhat shaky contraption a bit more acceptable to his horses, he carefully camouflaged the cracks with a layer of loose slough grass piled on top. Then he retired to his cabin for a well-earned rest.

Some time later in the evening, a rancher returning from a visit to a neighbor's place drove his team and buggy out on the bridge. Suddenly all four wheels dropped through the cracks, but his team was not inclined to stop on the shaky contraption and allow him to extricate the rig. They just kept going, breaking the double tree and running away, leaving the rancher and his wife marooned in the middle of the new bridge. When he saw that the bridge decking was lying the wrong way, the rancher began to fume, and he was muttering to himself as he went looking for some way to catch his team and get his buggy back on solid ground.

Coming to the Englishman's homestead, he knocked on the door and posed the problem, but he erred a bit in his approach by exclaiming to the proud engineer, "Some damn fool laid the decking the wrong way—"

He was rather taken aback when the homesteader interrupted angrily, "Damn fool am I! I say now, you can bloody well go catch your own team!"

It was not till the bridge builder learned that a lady was stranded in the buggy that he relented and came to the rescue, for these men, whatever their faults and shortcomings, were largely considerate and respectful where women were concerned.

Harold Butcher was no exception, and always enjoyed the good will and welcome of the ladies of the country, even if he was sometimes inclined to be a bit critical of the opposite sex. His criticisms were aired a bit unexpectedly at times. One incident in particular illustrates his eccentricity and dry wit.

The summer had been exceedingly hot and dry for weeks. The mountains and timber were like tinder, the creeks low, and the air had that parched, brassy feeling of a country that is thirsty and dusty.

One morning there was a smell of smoke in the air from a forest fire burning somewhere in the vast reaches of mountains to the west of us. At first this was only a hint of possible trouble, but as the days passed the smoke pall thickened until finally the sun was just a wicked-looking, blood-red disc hanging in the sky, and then it went out altogether. We found ourselves living in a kind of melancholy gloom that was like late evening in the middle of the day.

At that time any forest fires that broke out in British Columbia on the western side of the Continental Divide were largely ignored. With the exception of a few wandering trappers and prospectors, almost nobody lived in the region. It was a vast piece of country inaccessible except by pack horses in most places, so when a fire was set by lightning, it was simply allowed to burn itself out, for the British Columbia authorities didn't have much choice. Only when such a fire jumped the Divide to come down the eastern slope into Alberta on the edge of the range country would it be fought by big crews hired by the government.

On this particular occasion, we lived for several weeks in the midst of a cloud of smoke generated by a big fire in our neighboring province before we realized it was getting worse. One day about noon the smoke pall lifted, which is an ominous sign without a change of wind, for it means the draft at the fire has become strong enough to lift the smoke high in the sky in the immediate downwind side. By mid-afternoon there was a far-off, awesome roar like freight trains racing down a narrow mountain valley. It gave the impression of something mighty big gone berserk. Hour after hour this noise grew louder till it was a crescendo of thunder filling the mountains. When premature darkness came, we could see a great horseshoe of flames reaching across the mouth of Carpenter canyon from timberline to timberline on the south flank of Drywood Mountain. The draft from the heat of the flames was so intense that the trees ahead of the flames were suddenly cooked. The sudden steam thus generated blew the bark off them in twisted tendrils that were lifted high in the sky and dropped miles out on the ranches. Some of this bark picked up was scarcely scorched. This was more than just another forest fire eating up green timber—it was a fire storm racing miles in a few hours. A huge piece of heavily timbered valley up on the head of the south fork of Castle River about two miles wide and three miles long burned in an exploding holocaust of flames in about half an hour. The draft from this great flaming storm lifted burning embers miles over the range to the east, and so the fire came down Carpenter Creek toward the ranching country. It was about two and a half miles from our ranch when we first saw it.

Nobody in the country slept much that night. If the fire continued to run as it had been racing, many homesteads and ranches in its path would be in grave danger. Some of the people loaded personal belongings and family treasures in their wagons in preparation for flight, but there was little panic.

One homesteader—a sort of self-made and self-appointed minister of the gospel who particularly enjoyed peddling dire promises of brimstone and hellfire to all sinners—got more than his fill of looking at flames that night.

His cabin was on the brow of a bald-topped hill in a small

grove of aspens surrounded by stretches of well-grazed meadow land, and it was likely the safest location in the country. But for some reason, perhaps because they had a better view of it, they seemed to think that they were the only people in the country aware of the fire. About dark, they came rolling into the yard at our ranch with their team at a high gallop, and drew up wild-eyed to tell us to go quickly—the fire was almost on top of us. The man was black with soot, the woman was in tears, and the two youngsters were wailing in the midst of a heterogeneous mixture of belongings in the back of the wagon. We were somewhat at a loss for a reasonable explanation of their condition, but we managed to calm them, pointing out that the fire was dying down, as forest fires nearly always do at night. Mother fed them and we persuaded them to go home. Dad sent me along on my saddle horse to help them.

When we got to their cabin, I was treated to a sight that illustrated well what panic can do to people. Their home was a long, low, two-room affair with one door leading out onto a small porch at one end of it. The whole place was a complete shambles—about as thorough a wreck of a household as could possibly be accomplished even if it had been planned. From a somewhat jumbled explanation offered as we set about trying to put the place in a more livable condition, I was able to get some idea of what had happened.

As the fire approached with its rumble and roar during the afternoon this couple apparently fell to their knees and began to pray. Being highly emotional people with vivid imaginations, they must have imagined that the hell they talked about at Sunday gatherings was about to arrive. Their prayers did little to calm them, for neither had ever experienced anything to match this and it was understandable that they would be frightened. But when fear gave way to panic, the results were catastrophic, and a bit comical.

The man rushed out to hitch up his team to the wagon and then drove it up to the door, and while his wife and children wailed, he began loading some of their possessions.

The first thing he did was grab the huge cast-iron range, which must have weighed at least a quarter of a ton, and manhandle it into the doorway, where it stuck fast. He was a small man, and the fact that he was able to move the stove even that far was a superhuman feat of strength. But once stuck, it was immovable, so he grabbed the stove lids off the top and threw them into the wagon box. Then leaping back and forth over the stove, he carried a miscellaneous number of loose articles and bedding to the wagon. By the time this stuff was loaded it was black with soot. For every dish he managed to get into the wagon, there were more lying in shards all over the floor. The place looked as though a hurricane had struck it. Luckily there was no fire in the stove or the whole layout of buildings would have burned down.

I did what I could to help them straighten the place up into some semblance of order and then rode home.

All night long riders went by heading for the fire. Some people came through driving cars, which were just beginning to show up in that part of the country. Rumors flew in all directions. One had it that the canyons flanking Carpenter Creek to the north and south were also on fire. Another told of Mrs. Kootenai Brown, who was a Cree Indian and the widow of the famous frontiersman Kootenai Brown, the first superintendant of Waterton Lakes National Park; she had sealed herself in her small cabin down Yarrow Creek on the old Gladstone place, and was trying to bring on rain through Indian magic, or "making medicine" as it was called. No one laughed; we said nothing. If the Indian woman knew how to make it rain, we were all for it.

By dawn a crew of three hundred men was organized at the fire line being rapidly cut across the mouth of the valley through heavy growth of aspen and cottonwood on lower ground. Butcher was in the thick of this operation going full out to help the lone forestry official in charge. He appointed John and me to ride patrol up to the heads of two canyons coming down off the ranges to the north. It was our job to check the length of both of these and find out if any fires were burning in them. We were in the saddle and riding before it was fully light and covered over thirty miles of the roughest kind of country before noon. It was wild and spooky up among the peaks with the sun cut off by the smoke and the smell of burning timber in our noses.

It was shortly after noon when we rode back to the fire line to report. We had found no additional fires, and when we got back to Carpenter Creek canyon the fire there was burning low. It could not get going again when it came to the aspens, for this kind of growth is almost fireproof except in spring, when there are leaves and old dead grass under the trees. All efforts to set back fires upwind from the fire line had failed, and the crew was sitting in bunches here and there watching what was left of the fire when we arrived.

Shortly afterward there came a sudden darkening of the already gloomy day. Then a great rolling clap of thunder seemed to rip the sky into pieces, and the rain began to come down. It did more than just rain, it poured.

The weather had been dry so long scarcely anyone had even a coat tied on his saddle, and every slicker in the country was hanging on a nail at home. But we didn't care; we just climbed into our saddles and rode happily away with water running down our necks and out the tops of our boots.

Half a dozen of us rode with Butcher heading for his ranch and something to eat. I was riding beside him as we sloshed along the trail in the downpour. As usual he was bareheaded. The water was running off the end of his long nose, and he was staring straight ahead with a sort of bright-eyed look he got on occasion when something was getting under his skin. We had gone perhaps a mile, with the thunder and lightning going off all around us like big guns and playing a wild dance among the peaks, when Butcher gave a sudden snort.

"Just like a goddam woman!" he exclaimed. "Always overdoing it!"

For a moment I was at a loss to know what he was driving at, but then I remembered Mrs. Brown's prayers for rain, and whooped for joy. For a second or so he glared at me in utter disgust, but then he joined me in a great shout of laughter.


BUTCHER WAS LIKE no one else I ever met.

If he could comment on this remark, he would likely say, "Good thing! Bloody good thing! Be a frightful mess, what, if there were two of us!"

For certain sure, it would be interesting, and it might not be a bad thing at all, for there were not many human beings endowed with his quality of generosity. There was the time one of the neighbor boys was caught under a hay rack when it upset, and had an arm badly broken. The break did not heal properly, having been poorly set, and the arm went from bad to worse until the youngster was threatened with the loss of it. The family, like most of us in the hill country, was of limited means, and did not have the money to send the boy to a big city specialist, even if they had known where to send him.

Butcher heard of the youngster's trouble, and quickly went to the parents and arranged for the boy to be taken immediately to a bone specialist. It was all done quietly and unobtrusively, and in due course the boy was back, his arm completely healed and good as new. Very few people knew anything about what went on behind the scenes in this little drama. This suited Butcher, for he preferred to remain anonymous. Nor was this the only time he stepped in to help people in a quiet fashion. His neighbors owed him much.

The one thing he was selfish about was his library. Only on very special occasions would he lend anyone a book, and woe unto the person who failed to bring the book back in a reasonable length of time. But as a small boy I enjoyed a great privilege. From an early age I was an avid and enthusiastic reader, who could not find enough to read concerning adventure and travel around the world. Butcher saw me admiring his books one day when I was visiting at his ranch with my parents, and was somehow intrigued with this urchin who liked to read. He asked me what book I would take if I had my choice, and when I picked one of Kipling's, he was delighted. Quite by accident I had picked his favorite author. He went to the desk and got a small notebook and a pencil, which he tied on opposite ends of a piece of cord. He instructed me to write the book's title and the date in this, and then tied the string over a nail on the end of a shelf. It was, he told me, my tally book. Any time I wanted another book, I could come and get it. I was to write the title and date down, and when I brought it back, I was to draw a line through this record as a means of checking off the return.

As far as I know I was the only person to whom he extended this privilege. In frontier country, where books were not nearly as available as they are today through public and school libraries, this was a priceless opportunity, for it gave me a chance to read books that I would likely never see otherwise. Butcher, whether he ever came to realize it or not, ignited a small flame that never went out. I too owe him much.

Lionel Brook was of somewhat different stamp. He too was a highly educated man, but his attributes did not include very much real practical sense. He had owned two ranches, but for one reason or another had sold both. On occasion during periods of financial drought brought around by an absolute genius for throwing away money, he came to stay with Butcher, an arrangement that lasted about as long as it took for a check to come from England or for Butcher to run out of patience—depending on which condition arrived first. Then they parted, sometimes fiercely at odds for a time.

There was one occasion when Brook was at a very low financial ebb and was attempting to gather some tobacco money by trapping fur. Butcher came home from town one afternoon to find him skinning a mink on his spotless kitchen table. This was one of the times when Brook's welcome wore out completely before his quarterly stipend arrived, and he was forced to find lodgings elsewhere immediately.

Brook was largely unemployable and he depended mostly on the good will of his friends or his credit as a means of staying alive at such times. He was an accomplished artist in crayon pastels or oil paints, and sometimes painted pictures as a means of doing something for his keep. Many of the ranch homes had a mural painted by him on a convenient "beaver board" covered wall or a door panel. These were usually of mountain scenes with wild animals in the foreground. Because he rarely painted on anything but walls or doors, very little of his art survived, if any at all.

Brook was equally at home in a fine parlor or an Indian teepee. He could dress immaculately when occasion called for it, and being an accomplished raconteur, he was a welcome addition to any party. Complete in fine tweeds, polished boots, and pipe, he was the typical British sporting gentleman so often portrayed and lampooned in *Punch*. On occasion he would discard his thick-lensed spectacles for a monocle, and then, with his Van Dyke beard trimmed and brushed, he looked the picture of British aristocracy. With the monocle he was practically blind, but as someone remarked, he did not have to see to talk, and anyway, after a few drinks of Scotch he did not see too well even with his glasses.

On occasion he would don his greasy buckskins and go camp with the Indians. There he was particularly welcome, for he usually had money in his pockets.

Years after Brook was in his grave, I met one day an ancient Stoney Indian hunter away up on the head of the Highwood River between the Kaninaskis Range and the Rockies. When he found out I came from down in the Waterton country south of Pincher Creek, he asked me if I had known Brook. When I told him that I had, the old man's eyes kindled and he stood for a while silently looking far off at the mountains. Then squatting in his tattered buckskins with his head bowed and his long gray braids hanging down, he began to talk.

The old Indians had about three ways of talking all at

once. This man drew pictures in the dust with a little stick, made sign language with his hands, and filled in with broken English. I squatted on my boot heels in front of him, completely entranced by the picture he made and the memories he evoked. While the river murmured a few yards away and the breeze sang in the trees, we hunkered there with the great peaks looking down from all around, and his talk was of some good times gone by. The old man told me of hunting the high peaks for sheep, grizzly, and mountain goats with Brook, and how they rambled from one side of the Rockies to the other hunting and camping and just living under the big sky.

The old hunter finally wound up by saying with a certain nostalgia, "Him good man. Fifty dollars—one big ram. Fifty dollars—one goat. Much tobacco for hunters! Lots of grub for the squaw and papoose!"

With this I could heartily agree, for I too had fished and hunted with Brook as a boy. He was the first man to ever pay me for my services as a guide—a service I had proudly rendered at the ripe age of ten.

While wandering the creeks one day, I had found a hole full of bull trout—more correctly known as Dolly Vardens —the big chars found in western rivers closely related to the Eastern brook trout; and upon returning home found Brook visiting there. When I told him of my find, he immediately insisted that we go catch some of them, for no other fish are better eating. So I caught some large grasshoppers and put them in a tobacco tin for bait. This was a service Brook particularly appreciated, for the big chars will not usually take a fly, and his eyes were not sharp enough for catching grasshoppers in the long grass of mid-summer.

Upon reaching my fishing hole, Brook ensconced himself comfortably with his back to a tree, lit his pipe, baited his hook with a big grasshopper, and threw the unfortunate insect into the pool. Immediately a three-pound char took it and was in due course unceremoniously hauled out on the bank. I removed the brilliantly colored fish from his hook, rebaited it, and proceeded to take another.

When we arrived back at the house with a fine catch, he graciously gave the fish to Mother, and then reached for the beaded and fringed buckskin bag that he carried in a side pocket of his pants with the drawstrings looped on his belt. Opening it, he thrust in his hand and brought it out with all the silver coins it would hold, and put this fortune, the likes of which I had never dreamed of, on the kitchen table for me. My mother was utterly horrified at this kind of open-handed carelessness with money. It revolted her frugal Scottish heart to see money thrown around in such a light-hearted fashion. Furthermore, she was wise enough to know that it might not be good for a youngster to be so paid for just having a good time, so, somewhat testily, she made him put it all back but half a dollar; meanwhile I stood open-mouthed, watching the fortune that had come my way now disappearing just as fast. Brook was somewhat amazed and speechless at this unprecedented concern for his financial well-being, but he did as he was told with a few gusty harrumphs and snorts.

In spite of the fact that she treated him like a small boy on occasion, and sometimes bossed him unmercifully, Brook held Mother in highest esteem. He lavished compliments on her cooking, and bestowed on her all the respect and charming manner he would have given the cream of aristocracy. Naturally she loved it, and he was often invited to stop for a meal and spend an evening. Many the time we sat entranced as he unwound story after story of his adventures and travels around the world.

At various intervals one of Brook's wealthy relatives would die in England, leaving him a legacy. Then he would shake the dust out of his traveling clothes and disappear for months. But sooner or later he inevitably came back, for he always proclaimed that the Pincher Creek country of south-western Alberta was the most beautiful place in all the world, and he was not far from the truth.

Once he spent the winter in Hawaii content to loaf and play away the time under balmy Pacific skies. But one day he suddenly realized that it was greening-up time in the Alberta Rockies and the mountains would be exchanging their man-tles of snow for the new green of grass and the brilliance of a myriad flowers. Heeding his itchy feet, eager to be on the move, he caught the first boat out of Honolulu and sailed for San Francisco. Upon arriving there, he walked off the ship up onto Market Street to hail a cab.

Getting into the taxi, he instructed the driver, "Take me to Pincher Creek."

"Where is Pincher Creek?" asked the driver.

"Up in Alberta," replied Brook somewhat testily.

"Excuse me, sir," said the driver with typical American ignorance of the neighboring country, "but where the hell is Alberta?"

"By Jove, my good man," Brook exclaimed. "It's up in Canada!"

The taxi driver was somewhat astonished, to say the least, when Brook finally got it across to him that he was expected to drive fifteen hundred miles north into the wild hinterland of a foreign country; but he was game. He drove to his house, packed his bag, kissed his wife goodbye, and they set forth. In those days the roads were sketchy and adventurous; but some two weeks later, having braved snowy mountain passes, flooded rivers, searing deserts, mud, and innumerable flat tires, the driver finally brought his mud-splattered vehicle down the main street of Pincher Creek.

Their first stop was the bank, where Brook withdrew eight hundred dollars in cash to pay his taxi fare. Then he took the driver down the street to the general store, where he had him fitted out with the finest suit, handmade boots, and ten-gallon Stetson hat that money could buy. Having bestowed these gifts on the taxi driver, they both got all dressed up in their finest and drove on a kind of grand tour of the neighborhood to greet all of Brook's friends. This took several days and included some memorable parties. In due course, the taxi driver turned his car south, heading for California—no doubt a bit better informed about Canada and perhaps somewhat astonished at the kind of people who lived there.

OUR RANCH WAS LOCATED in the midst of rolling green hills among parklands of aspens, cottonwoods, and willows, where dozens of small lakes and sloughs glistened in the sun. The country was full of game, big and small. There were sheep, mountain goats, elk, deer, and bear in the mountains. The tracks of furbearers were everywhere. Sharptails jumped in flocks from the meadows and the bush was full of ruffed grouse. The whole region lay under the great mountain flyway—the migratory road of the wild fowl—and was a great nesting ground for these birds. It was here that my brother and I got our first taste of hunting.

For many years Brook and our grandfather rendezvoused at the ranch to join Dad in the annual opening of the duck season. John and I were still too young to use guns, so we carried ammunition and dead birds, and shook with excitement at being included in the hunting amidst the banging of the shotguns. We got delightfully exhausted and muddy helping the black Labrador retriever that Brook borrowed each year from Butcher.

His name was Mogs, and he was just as eccentric as his master in a different kind of way, for he had habits unlike those of any other dog I have ever known. His appetite and taste were beyond explanation or reason, for in spite of the fact that he was pampered and overfed to the point of obesity, he would eat anything at any time. He once sneaked up behind me when I was fishing and before I noticed him he quickly gulped my entire catch of trout. There was another time when he caught and ate a huge toad, whereupon he got horribly sick in the back of my grandfather's touring car. Most dogs hate the taste of duck—some so detest it that they have to be force-trained to retrieve them, and even if the duck is cooked, most dogs will not eat it unless at the point of starvation. But Mogs was different. First thing each hunting day we shot a mudhen for him, which he was encouraged to eat; otherwise he would swim out and get the first duck and then stand just out of reach in shallow water a few steps from shore while methodically chewing the bird to rags. Then he would swallow it in great gulps, feathers and all. He insisted on eating the first bird, and to save a good duck, we arranged for it to be a mudhen.

Unlike most dogs, which are completely cured for life by one painful contact with a porcupine, Mogs had a mania for chewing on the quilly beasts. He would go out of his way to find one, and would inevitably get well quilled. Then he would come and sit stoically, whimpering and crying, while the quills were painfully removed. Knowing this masochistic fascination of Mogs's, no one acquainted with him ever went out without a pair of pliers in a pants pocket. Hunting with Mogs was never dull, nor did it tend to run in any kind of pattern. Anything could happen and quite often did.

On those early fall mornings of the shooting season we set forth before sunrise, when the smell of dew on ripening grass was spicy in the air. Frost sometimes sparkled in the low places. And the first hint of gold in the willows and aspens was visible. At times the lakes and sloughs were all smoky-silver with a thin blanket of fog lying on top the water. Hidden in it were the ducks, and we knew they were there even if they were out of sight, because we could hear them gabbling and chuckling over their morning feed.

We were a motley crew: Grandfather with his silvery fringe of hair showing around his hat and the inevitable pipe throwing off clouds of smoke; Brooks stalking gingerly through the wet grass, his legs spidery in tight-fitting breeches, looking for all the world as though he had just stepped out of the pages of the *London Illustrated News* or *Scottish Field* with his tweed shooting jacket and cloth cap; and Dad striding along with his shotgun at trail position. Bringing up the rear were the dog and two bright-eyed urchins, their pants pockets bulging with extra shells, generally shivering a bit with the cold and the anticipation.

As the sun climbed over the far eastern rim of the world, red and warm, each tiny dewdrop on its setting of grass blade or twig was lit up like a glittering jewel. Then we would come up into the willows fringing a slough, where inevitably ducks leaped into flight in front of us. Teal, pintails, and mallards were our game. The shotguns lifted and began their ragged hammering; birds would fall—suddenly stilled in mid-air as though jerked with strings—and there would be the sharp, exciting smell of burned nitro in the damp air.

So the long mornings went until we were of a mind to

quit for the simple lack of desire to kill more ducks. By noon we came trailing home through the sun-drenched groves of aspens bearing our feathered plunder, famished to the point of aching, and filled with the sheer primitive joy of returning with meat for the roasting fire.

Nothing—not even the feathers—was wasted. We ate roast duck till we threatened to burst. Mogs ate what was left over. Mother even saved the feathers for stuffing pillows and bed comforters. John and I slumbered with the soft warm smell of duck feathers, for we slept on a feather mattress the year round.

We enjoyed that first hunt each year more than anything I can remember. It was more than just a hunt; it became a sort of ritual—a rendezvous where we enjoyed each other's company, exchanged man talk, and dipped into the purely atavistic desire to prey for the sake of enjoying good food. The urchins grew into lanky, ropy-muscled boys, but the years piled up on Grandfather and Brook, slowing their steps, though not dulling their love for the sport. Grandfather's shoulders became so stiff that he had difficulty making that smooth swing so necessary in the wing shooting he loved so much. Brook's eyes, never the best, became dimmer and dimmer. One frosty September morning we were suddenly aware that he was shooting at the sound of ducks he could no longer see except as shifting shadows. I had graduated to a shotgun by that time, and along with Dad shot birds that Brook was struggling to hit, and as well as we could we timed our shooting so that he would think the birds were his.

We thought we were getting away with the deception, but when noon came and we were all sitting on the back steps of the house admiring our morning's bag, we noticed Brook was not his usual boyish self. He sat quietly looking down at the fine old English sidehammer shotgun across his knees. Like its owner it was old; its rich engraving revealed the skilled hand work that had crafted it, and the fine walnut stock glowed with the rubbing of loving hands.

As though coming to a decision, Brook suddenly lifted his head and flashed the grin we knew so well.

"Boys," he said, "you have been good to an old man, but I am through with hunting." Then handing his shotgun to John, he added, "By Jove! You need a gun. This one is too good to end up hanging on pegs with nothing to do. With your father's permission I give it to you, and hope you enjoy it as much as I have."

For a long moment we sat there very quiet, and then Dad nodded, got to his feet, coughed, and went into the house. John and I suddenly found ourselves sitting alone with old Mogs, the ducks, and our guns. We scrubbed at our eyes not looking at each other, but the tears continued to roll down off our cheeks in spite of the scrubbing. Somehow we knew right then that a wonderful part of our lives was gone for good, and with the knowledge of its going we were for the first time aware that our lives were just as fragile and uncertain as those of the birds we hunted. We had felt one of the first cruel pangs of growing up.

We enjoyed many more good hunts, but somehow none was ever quite the same. But the memory of these first mornings is good—something revived in the clarity of crystal dewdrops seen on chilly September dawns, when the sun rolls up over the far rim of the world, red and warm, to light a new day and hail the coming of another hunt.

6

# The Wilderness Fisherman

As a boy of a size "somewhere between weanin' and chewin' tobaccer," as Charlie Wise, an old friend and mountain trapper, would put it, I ran the gamut between tickling trout, noosing fish with a loop of snare wire tied to the end of a stick, throwing a bait to them big enough to keep a good-sized cat fed for a day, presenting an impaled grasshopper on a snelled hook, and angling with hand-crafted artificial flies so small it was difficult to see them well enough to tie them on the hair-like tippet of the leader.

To those who have never heard of tickling trout, I will explain: it is an old English and Scottish poachers' trick, which I learned from reading a classic entitled *Lorna Doone* and from the further instruction of our old friend Lionel Brook. It is only possible in small streams or in such places where one can reach into a hide with a hand. By the use of much patience and practice it is possible to slip a hand under a fish and stroke its belly, which lulls it into allowing the fingers to be slipped under the gill covers and clamped shut. The fish is then thrown out of the water onto the bank. I might add that very few people have the patience to master the muscle control necessary for success. On one or two occasions when I was hungry and had nothing much to work with but my hands, I used this method with success. One time I took a fair-sized Dolly Varden out from under an overhanging ledge and cooked him on a flat rock over a fire without anything else to garnish the result. My father and I ate well off that fish and enjoyed what would have otherwise been a very hungry ride.

Snaring fish is largely illegal and justifiable now only as a means of survival. All one needs is a piece of brass or copper wire from fourteen to eighteen inches long. For small fish only one strand is necessary, but for larger kinds this may be too light and then two or three strands must be twisted or braided together. The snare is tied on the end of a light pole with the loop open. It is a matter then of just carefully slipping the snare over the fish's head, jerking it shut just back of the gills, and swinging the captive out on the bank in the same motion. Of course this can be done only in clear water where fish are within reach. In the smaller streams along the mountains in summer we took whitefish in this manner, when food was the first consideration.

When someone came who was likely to stay for a meal,

my mother, upon finding her pantry a bit short of something to serve quickly, would send my brother and me to the creek for some fish. In half an hour we would usually be back with all the whitefish needed. Just to see if it could be done, I once snared a ten-pound great northern pike with an ordinary braided cotton shoelace tied on a forked willow. It is a good way to get a meal in wilderness country if one has no better way, but there is absolutely nothing sportsmanlike about it.

The early settlers along the foot of the Rockies did not use nets. A net is expensive to make or buy, and this means of taking fish was largely unnecessary, for few people needed or wanted that much fish. Dynamiting was another means of taking fish practiced by a very few. It was frowned upon by most everybody, for it destroyed everything in a section of stream and was extremely wasteful. On rare occasions dynamite is still used by unscrupulous types. The only time in my life I was ever tempted to throw down on a man with a gun was when I caught three coal miners dynamiting trout.

We were coming back through the Rockies with the packtrain in August, after a camping trip on the Flathead River in southeast British Columbia, when I rode back to an overnight camp alone to get my spurs, which had been left hanging in a tree. That morning we had passed three coal miners on a fishing holiday hiking down the trail with packs on their backs. Where the trail rimmed a deep gorge not far from our campsite, I heard the unmistakable boom of a blast from down in the bottom. When I left my horse and climbed down to investigate, I came out of a fold in the cliff face within ten steps of this trio of tough-looking characters all busy retrieving fish killed by the blast. At their feet among the rocks were several large Dolly Varden trout. Upon my somewhat smoky query as to what they thought they were pulling off, they turned in momentary astonishment. But this did not last long. Two of them started toward me with that flat, stone-hard look men get in their eyes when they are closing in on one of their own kind with nothing good in mind. I was wearing a long-waisted, heavily fringed, beaded moosehide jacket, which somewhat camouflaged the gun belt and holster strapped around my hips. In the holster was one of old Sam Colt's historic "equalizers"—a famed .44 caliber single-action sixshooter, and at that moment I was very glad to have it within reach. I didn't touch it, however, but just peeled back my coat out of the way and hooked my thumbs in the belt. The atmosphere and general climate around there changed very fast, and the gentlemen in front of me acquired the look of those who were all set to club a rabbit and suddenly found they were declaring war on a grizzly bear. They did not try to run, but melted down to size while I read them the riot act. Had I not been over a hundred miles from the nearest conservation officer and in no position to lay charges, I would have been delighted to turn them in for prosecution. As it was I hazed three chastened men back up the trail and escorted them out of the province of British Columbia with a solemn promise of what would happen if they chose to come back. I never saw them again.

I carried that pistol for years, with a special permit to make it legal, putting up with its weight and inconvenience with the thought that some day I might need it to discourage a bear that might decide to chew on me or kill a horse if someone got a foot hung up in a stirrup. In all the thousands of miles of wilderness country it was never needed except on that day, and, strangely enough, to protect myself from some of my own species. Had I not been carrying it, I might have bluffed or fought my way out of a tough spot, but likely not without some considerable expense one way or another. As it was, the episode amounted to very little, nobody got hurt, and perhaps some good was done—all of which may boil down to some kind of moral in the context of some modern-day questions.

WHEN I WAS STILL little more than knee high, I can remember my father putting a set-line in the St. Mary's River to take fish for the table. A set-line, or trot line as it is called farther south, was generally anchored on each shore and adjusted so that the length of the line lay across the bottom of the river. At regular intervals a shorter piece was tied to this line—each with a big hook on the end. The hooks were baited with a piece of raw meat or a small fish. Such a line was usually attended with a boat, and with it a considerable number of fish could be taken at once. But where no boat was available, the trot line had a heavy weight attached to one end for an anchor; the other end was attached to a stake or tree limb along the shore. This was the kind my father and mother used for northern pike, walleyes, and other fish found in the river. There were very few trout in the lower reaches of the St. Mary's, but plenty of coarse fish, including the occasional enormous sturgeon, Lake Winnipeg goldeyes, ling, or burbot, along with millions of various kinds of small fish.

It was always something of an adventure to haul in a set-line. Mother would shriek with excitement as she pulled in a malevolent-looking monster of a pike with a mean gleam in his eyes and a set of teeth that would do justice to a small alligator.

The pike is a predatory fish looking something like a fresh-water variety of barracuda. It will eat about anything small enough to get down its throat, and has been known on occasion to choke on something too large to fit. Small fish are its usual prey, but it will also take small animals and birds if the opportunity arises. I once saw a mother mallard moving her brood from a high-ground nesting site to the Waterton River, up near the place where it flows from Waterton Lakes. When she came to the river with her brood, she headed for an island in mid-stream, but before she had gone more than a few feet, a V-shaped wave headed her way at high speed from one side. There was a sudden heavy swirl and a duckling straggling a bit behind the small flotilla was suddenly not there any more. At first the mother duck was unaware of the deadly menace stalking her family, but when another duckling disappeared, she sped into a flurrying scramble. The surviving young ducks almost ran on the surface in their efforts to keep up to her, but in spite of her efforts to get her family out of reach of the big fish, half the brood was gone by the time she reached the shelter of the island.

I once dressed a northern pike weighing perhaps five or six pounds that had just eaten a young muskrat before it took my spoon. Another time I found a half-grown ground squirrel in the stomach of one.

At Grandfather's ranch one afternoon following a heavy thunderstorm, I found a barn swallow's nest knocked down from under the eaves of one of the buildings. There were four almost fully grown and fledged young swallows lying drowned in the wreckage of the mud nest. Not really knowing why, I picked up the dead birds and carried them down to the river, where I casually tossed one into the water. The heavy current swept it down around the end of a submerged reef of coal projecting from the bank, where there was a sudden electrifying explosion from the depths and the bird disappeared. The next swallow that went floating along the same route had a fairly substantial hook impaled in it, which was bent to a wire leader tied to a stout braided cotton line. My rod was a peeled willow, to which I clung with both hands with the butt braced against my belly.

There was still a lot of slack line coiled in loops and bends on the surface when there was a sudden flash of a big fish and the water flew as the floating bird disappeared. The line went shooting out toward the middle of the river at high speed, and there was barely time to get my legs braced when a huge pike ran into the end of the slack like a suddenly snubbed bronco at the end of a rope. Up out of the water the big fish jumped in a great tail-snapping leap, ending up in a big splash as the tight line jerked it down. The pike was hooked deep in the gullet, and when I finally hauled it out on the sand bar and quieted it with a stout club, it appeared even bigger than it had in the river.

It was a handsome fish in a tigerish sort of way and all the load a ten-year-old could manage. When I picked it up by a gill cover, its nose came up even with my shoulder while its tail still dragged on the ground. It was never weighed, but over the intervening years it still lingers sharply in my memory as being one of the biggest fish I ever caught. It was about the closest I ever came to taking pike on a dry fly—although many times some good catches have been taken on a floating plug made of wood or plastic—all painted in gay colors and armed with treble hooks. Even though these pike were taken on light tackle that would have made my peeled pole of that day on the St. Mary's look like a telephone pole by comparison, that battle stands out as being a combat of sheer strength between a small boy and a big fish—primitive, fierce, and unforgettable.

I once knew a trapper who in the spring tended a line by canoe along a river collecting beaver and muskrats. It was a wild, uninhabited piece of country, and his only contact with the outside world was a small village upstream a few miles, where a railroad crossed the stream.

Upon running short of some supplies, he paddled up to this place one day to replenish his stock. Along toward evening he headed back downriver, loafing along with the current, carrying his canoe silently on its way. It was a very pleasant, balmy evening and the man was enjoying a good cigar as the canoe slipped past groves of silent spruces and cottonwoods under a rich canopy of stars studding the night sky. Taking the cigar from his mouth he knocked the ash off into the water, exposing the glowing coal on its lighted tip. Before he could put it back in his mouth, a great pike came lancing up out of the river and closed its teeth on his hand. He threw the fish in the canoe and managed to extricate his fingers from its mouth, but not before suffering bad lacerations. He went on to his camp, but the wound somehow got infected and did not heal for a long time. This man still carries the scars to prove that pike sometimes snap at anything when hungry.

The pike is generally not considered to be of top eating quality; but if carefully filleted and skinned, then deboned by pulling the forked sidebones out of the fillets with a pair of pliers, it is delicious when caught in cold water. Its flavor can be greatly enhanced by sprinkling the raw sides of the fillets with a teaspoon of rye whisky as they fry.

The walleyes rarely achieved a weight of more than six or seven pounds. These were delicious fresh from the river fried in bacon grease. John and I had our best luck with them as the water cleared after a heavy rain—especially following the spring flood. If one was caught, there always seemed to be more in the same hole, for they ran in schools and fed like hungry wolves.

# THE WILDERNESS FISHERMAN

Almost every time we went fishing in the St. Mary's, we caught one or two Lake Winnipeg goldeyes, a Canadian fish found nowhere else but in Lake Winnipeg and the Saskatchewan drainage. The goldeye is a pretty fish, with large silver scales and bright gold eyes. It grows up to sixteen and seventeen inches in length and sometimes attains a weight of two and a half pounds. We occasionally caught them on the surface with a floating grasshopper, but more often with a polished copper Colorado spinner or a small minnow while fishing for pike or walleyes. The goldeye is a tasty panfish when cooked fresh, but reaches gourmet heights when smoked.

The first time I ever set a line in the river all night I hooked a fish I had never seen before. When I went early in the morning to haul in the line there was something heavy and sluggish at the other end. Bringing in the line revealed a great ugly fish, almost as long as I, with a slippery smooth scaleless hide colored in yellow-green mottles against a lead gray background. Its tail was rounded instead of forked, as were all its ventral and pectoral fins. Running most of the length of its back was a long, ragged dorsal fin like a piece of wet limp leather. I had captured a monster of a ling with a great pot belly and a mouth like a cavern.

The ling, or more correctly the burbot, is a kind of fresh-water cod. It looks like something held over from another era in the evolution of life on earth—a time when strange and wonderful animals wandered among the steaming swamps—and quite likely it is. By trout standards of good looks, it is a most unappetizing-looking fish, and more than one modern-day bait fisherman has thrown such a catch away. Nothing could be more wasteful, for the ling, upon being skinned, filleted, and properly cooked in bacon fat, is a most delicious fish, almost indistinguishable from fresh cod; its meat is sweet, snow-white, and rich, with only a few large bones.

Although the huge sturgeon was a very rare catch on the far western reaches of the Saskatchewan drainage, every so often a set-line fisherman would catch one of these fish. I recall a coal miner hooking and landing one on the St. Mary's that was about as long as himself and weighed in the neighborhood of seventy pounds. As sturgeon go it was only a baby, but a monster of a fish just the same, which warranted headlines in the local newspaper and gave the fisherman a measure of fame for miles around.

FOOD CAME FIRST and sport second among the early settlers where catching fish was concerned. The rivers and creeks teemed with fish then, and while there was no kind of refrigeration to keep fish for very long, smoking and salting were the methods employed to preserve fish for winter use. Some of the ranch women also canned fish in Mason jars with a light vinegar sauce, which softened the bones and kept the meat perfectly for a long time. For the most part fish were eaten fresh.

Up in the streams under the face of the Rockies we found a different world, where my brother and I not only enjoyed the ultimate in piscatorial adventure, but also learned to appreciate and practice sporting methods of taking fish.

Down either side of the great rugged mass of Drywood Mountain the north and south forks of the Drywood flowed clear and sparkling cold to join on our ranch. We called the North Fork Butcher's Creek and the other Carpenter Creek, after two ranchers who lived along the banks, while below the forks it was the Drywood as designated on the map.

These creeks flowed over brilliantly colored rocks of every shade, where sun and shadow played in the swirling currents, and cutthroat trout, whitefish, and great Dolly Vardens lurked and fed—vastly exciting, undulating shadows moving under graceful overhanging willows, aspens, and cotton-woods.

Later the rainbow trout was introduced, and we found the added thrill of fighting a great sport fish, which leaps high out of the water when hooked.

There was a smaller stream—the middle fork of the Dry-wood—that flowed down from between the shoulders of the mountain through miles of aspen groves, and this was a beaver haven, with chains of dams providing small boys with fishing grounds like something dreamed up in paradise. These ponds were stiff with cutthroat trout, so many they rarely grew beyond half a pound in weight, and so eager for food we saw them sometimes strike at a bare hook. Once I remember taking a good catch with aster blooms for a lure and another time with fragments of puffball. Mostly we used grasshoppers for bait at first, and stalking trout in the bushy, tree-bordered dams was a kind of careful hunting for individual fish hidden in innumerable hides. Mere catching of fish soon palled, so we laced our angling with the excitement of seeking and catching the biggest ones. We also progressed from bait fishing to taking fish with a fly.

Our first rods were crude, the traditional willow poles so often portrayed as the choice of small boys. These were heavy, awkward, and not much fun to wield after an hour or two of fishing. We substituted light cane poles—eight or ten feet cut from the tip of a binder whip shaft used when these harvesting machines were drawn by horses. Such a pole cost about fifty cents and was a vast improvement over a crooked willow. Then we progressed to the lighter and handier telescopic and jointed-steel bait rods then sold by every hardware store and mail order house. The best one I ever owned cost ninety cents at Eatons—a Canadian mail order house equivalent to Sears Roebuck. It was light, whippy, and strong; it collapsed into three sections for carrying and lasted a long time. We also used steel telescopic bait rods with sections that pushed down into each other, so the rod could be more easily carried through the bush. But most of these rods had the habit of either jamming or coming apart, so they usually ended up stuck or soldered at full length. When carrying such a rod we usually just wound the line up on the reel until the hook caught in the tip guide, and this practice once caused an explosion of unexpected action in a most unusual way.

We had spent the day fishing Carpenter canyon with some neighbor boys and were riding down the valley in late afternoon, our saddles hung with bulging fish sacks and our rods in our hands. One of the boys was riding a big, lazy, sleepy-looking white horse that had a habit of poking along as though reluctant to put one foot ahead of the other, his tail at half-cock, and breaking wind at every other step or so. He was about as inelegant as a horse can get. At one point along the winding trail a fly lit on his big, ugly nose. He stopped abruptly to wipe it off on his front leg, and at this point the rider coming behind, also half asleep, ran the tip of his rod up under the horse's tail, where the hook caught on the edge of his equine exhaust pipe.

The results bordered on sheer magic. That old white horse came alive with a great snort and a thunderous sound of exhaust as he went straight in the air like a three-star rodeo bronc. The reel on the surprised fisherman's rod screamed as line tore off it, and that startled individual helped the uproar into a resounding crescendo by rearing back on his rod as though fighting a leaping tarpon. About the third jump the white horse threw his rider into the tops of the trees along with his sack of trout, and then ran away, still impaled by the hook and trailing a length of broken line.

It took the three of us a good half hour to run down that horse. Several times we managed to get close, but somebody's horse invariably stepped on the line, which launched another stampede for distant places. When we finally managed to get hold of him, we were still faced with something of a dilemma, for the hook was still imbedded in a very tender part of his anatomy and trying to remove it was a barefaced invitation to get your head kicked off. Somebody hobbled him with a halter shank, and then we tied up a hind foot with a lariat so he was standing on three legs. A jacket was employed as a blindfold to keep him in the dark. Then some crude surgery was carried out with a pocket knife without benefit of anesthetic. The operation was successful, but ever after that when we went fishing with the boy who rode this horse and he began to poke along holding up the line, all someone had to do was pull a little line off a reel behind him. At the sound of it, the lazy white horse would tuck his tail down and take off like a scalded cat. He was henceforth endowed with what might be termed a conditioned reflex.

SOMETIME AROUND 300 B.C. the first reference to "luring fish with feathers" was made. Likely the first artificial lures of this kind were minnow imitations used in jigging or trolling in salt water by fishermen in the Red Sea and Mediterranean. Another historic reference cites the third century A.D., when a naturalist by the name of Aelian wrote about fly fishing in his scroll *De Natura Animalium.*

In a chapter entitled "De Peculiari Quadam Pisatu in

Macedonia," he says: "There is a river called Astraeus flowing midway between Berea and Thessalonica, in which are produced certain spotted fish whose food consists of insects which fly about the river. These insects are dissimilar to all other kinds found elsewhere; they are not like wasps, nor would one naturally compare them with flies called ephemera, nor do they resemble bees. But they are impudent as flies, as large as Ant Hedon, of the same color as wasps and they buzz like bees. The natives call this insect the 'Hippurus.'

"As these flies float on top of the water in pursuit of food, they attract the notice of the fish, which swim upon them. When the fish spies one of these insects on top of the water, it swims quietly underneath it, taking care not to agitate the surface, lest it should scare away the prey; so approaching it, as it were, under the shadow it opens its mouth and gulps it down, just as a wolf seizes a sheep, or an eagle a goose, and having done this it swims away beneath the ripple.

"The fishermen are aware of all this; but they do not use these flies for bait because handling would destroy their natural color, injure the wings, and spoil them as a lure. On this account the natural insect is in ill repute with the fishermen, who cannot make use of it. They manage to circumvent the fish, however, by the following clever piscatorial device. They cover a hook with red wool, and upon this they fasten two feathers of a waxy appearance, which grow under a cock's wattles, they have a reed about six feet long and a line about the same length; they drop this lure upon the water and the fish being attracted by the color becomes extremely excited, proceeds to meet it, anticipating from its beautiful appearance a most delicious repast; but as with extended mouth it seizes the lure, it is held fast by the hook, and being captured, meets with a very sorry entertainment."

So goes what is likely the first historical reference to taking fish with artificial flies—a very graphic and revealing description of an insect and the means of copying it in order to take trout. Sport fishing was very likely born this way, for no man could fish with such delicate tackle without some practiced skill; nor could he see fish coming up through clear water for his hand-tied artificial fly without some feeling of meeting a wild thing on its own terms and enjoying the contact as much for the experience as the possible kill.

SPORT FISHING ON the Alberta frontier along the tributaries of the Saskatchewan was something that grew on the people through contact with English remittance men and visitors from those parts of the world where a certain protocol had been developed in the sports of fishing and hunting by the necessity of conservation and population pressures.

Several centuries ago, British landowners had set up certain rules and sporting traditions in the development of game management in that country, which was second only to the system practiced in Germany at that time. Over the intervening years Germany still maintains the lead, for their refinements of sporting manners, traditional rules, and protocol afield and along their streams are the ultimate. Here in North America we too have come a considerable distance in a comparatively short period of time, but we still have a long journey ahead before we can match our management programs with those used by European biologists and foresters. North American authorities defend our system by saying that the same kind of program as used in Europe will not fit here, which is true to a point, but the framework of principle and knowledge involved will hold anywhere. To a large extent our government fish and wildlife departments are more interested in the harvest and the resulting income from sale of licenses to sportsmen for the privilege of hunting and fishing than they are to the development of true appreciation of sport and safety. They pay lip service to stream pollution problems, but do nothing to really get at the roots of the trouble. If a license purchaser does not know the difference between a rifle and a shotgun, he still has the legal right to go hunting.

THE FIRST ARTIFICIAL TROUT FLY I ever used was a snelled Royal Coachman wet fly, bought at the local hardware store more because it was very pretty than because I thought it could take trout. But the trout of my favorite beaver pond went hog wild over it, chewing it to rags in no time, but not before I had taken plenty of the biggest ones for a feed for the whole family. From then on I was a confirmed fly fisherman, although my steel rod was anything but delicate and my money supply did not include enough for a stock of store-bought flies. To solve this deficiency, for my first homemade fly, I pulled some feathers out of an outraged Plymouth Rock rooster and proceeded to tie something out of red wool, Christmas tree tinsel, and ordinary sewing thread. It was born looking like nothing ever seriously contemplated by a purist and made a splash like a drowning chicken. But, amazingly enough, it caught trout. As a matter of fact, the first time I used it, my heart missed a beat when a great Dolly Varden struck it hard, no doubt mistaking it for a small bird that had fallen into the creek. The big fish smashed into it in clear water within two feet of my toes as

I stood on a low undercut bank rimming a deep pool, and I came within a whisker of falling in over my head. Thus a fly fisherman was born.

The first really good fly tackle I ever saw belonged to a judge of the circuit court, who spent his summer holiday for years camped by the creeks on our ranch or the adjoining one belonging to Butcher. The Judge, as everyone called him, was a Scotsman by descent, with a great love of camping alone in the midst of the wilds and a taste for whisky. My first acquaintance with him got off to a very poor start, although my intentions were good and I very likely saved his life in the process even if it was crudely done.

During July and August the big Dolly Vardens came up the mountain creeks from the rivers; their September spawning grounds were along the bars and riffles near the headwaters. Hunting and catching these beautifully colored big fish was a never-ending source of excitement, and they were wonderful eating besides. So every evening when I rode down into the valley to bring in the milk cows, I went along the banks of the creek looking down into the pools to spot these fish as they lay on the bottom. The Dolly Varden, like its very near relative, the eastern brook trout, has a distinctive and sharply contrasting border of ivory-white along the leading edges of its pectoral and ventral fins, a color that often gives it away to sharp eyes even when hidden by logs or overhanging ledges. If I located one or two of the big fish, I came back early the next morning properly armed for their capture.

One particular evening I came to a big pool at the forks, and as I rode up on a gravel bar dividing the creeks at their junction a surprising sight confronted me. There in the middle of a fast riffle dropping into the pool sat a somewhat portly gentleman in long waders. The waders were awash, for the creek was running cheerfully over the top of them at the back, and the man was in grave danger of being swept down into deep water, where he would doubtless sink like a stone. It was the Judge.

For a moment or two I just sat there on my horse in astonishment, not fully aware that he was in real trouble. But when he turned to look at me, his expression gave away his fear, and it became instantly obvious that he was pinned down by the weight of water, virtually unable to move.

I was only a kid weighing perhaps ninety pounds, while the Judge in his present predicament with his waders full of water, likely weighed two hundred fifty. To attempt to drag him out, as one would normally do, was just asking for trouble. Had I been afoot, the problem would have been near insurmountable; but I was astride a wiry little cow horse with a great love of snapping things on the end of a rope. Just as naturally as one would reach for his hat upon entering someone else's house, my hand dropped to the coiled lariat hanging on its strap on the fork of my saddle. Before the Judge was aware of my intent, the mare came splashing out toward him and a loop sailed out to drop and be jerked up snug around his arms and chest. In the same motion I dallied the rope around the saddle horn, and then with a great deal more enthusiasm than diplomacy my mount spun on her heels to head for dry ground. The Judge came sliding out of the creek backwards, somehow managing to hold his rod out of harm's way, and when I checked my horse he came to a stop head down in a hollow on the gravel bar. He was instantly inundated by a rushing flood issuing from his waders. He came up on his feet muttering thick Scottish words of import not found in courts or churches as he shook off the rope. My horse took one horrified look at him and bolted, which was probably just as well, for at that moment

I doubt very much if the Judge was entirely cognizant of my good intentions.

The Judge's fine split-cane English fly rod had survived the action unscathed, but a few days later it met an incongruous end.

My friend Butcher, in a burst of enthusiastic extravagance that summer, had purchased a brand-new, magnificently appointed McLaughlin-Buick sedan—a grand car with an arrogant profile to its general outline and a powerful engine under its long bonnet. It was a favorite model of rum runners during prohibition, a famed "Whisky Six," a fast, rugged machine that would stand up to use on the sketchy roads of the times. It had one idiosyncrasy of design—the reverse-gear position was that of low gear in most other cars; so new owners who had learned to drive ordinary ones sometimes found themselves going backward instead of forward as expected.

Naturally Butcher was delighted with this new toy and drove it into our yard to show it off. Only my mother and I were at home, but we were a satisfactorily appreciative audience as he explained and pointed out all the finer points of his new car.

Finally Butcher turned to me, "Come along and open the gates for me. We'll go visit the Judge."

So I sat on the big leather-upholstered front seat beside him and away we went down a twisting wagon trail into the valley where the Judge had his tent pitched. The Judge was warm in his welcome and very interested in Butcher's new car. He uncorked a full bottle of whisky to celebrate the occasion as he and Butcher exchanged the latest accounts of happenings in the country well laced with humorous comments. Being too young, I was not invited to join them in a drink, but I sat enthralled as they laughed and chuckled at various stories. As the talk proceeded, the bottle was passed freely back and forth, its potion being slightly diluted with creek water in large glasses. By the time Butcher was ready to leave for home, he was feeling no pain.

He shook the Judge's hand with decorum and settled himself with immense dignity behind the steering wheel of his car. The door banged shut as the starter whirred and the engine caught in a throaty purring of power that spoke of many horses waiting to be turned loose. Butcher threw the shift lever out of neutral into gear and let up on the clutch pedal, whereupon we began to go backwards.

The Judge had been standing by the car as this happened, and he came trotting along beside the open window to pronounce with a certain judicial firmness of tone, "But Mr. Butcher, you are going backwards!"

"Nonsense, Judge," Butcher replied, "I am in low gear."

About the time the Judge opened his mouth for some further comment on Butcher's mechanical misconceptions and direction, there was a vast clatter of various things, a ripping of canvas, and some lurching as the car went over a six-foot bank to land with a great crash in the creek. The engine stalled and Butcher opened the door to get out, but he changed his mind when he saw cold water lapping the floorboards of his vehicle. Somewhat sobered but still firmly in command of things in general, he started the motor again and drove straight down the bed of the stream to a low place in the bank where the trail crossed. There he swung the car back up on the beach and returned to what was left of the Judge's snug camp.

Together he and the Judge stood solemnly surveying the wreckage. The tent was flat and torn, the sheet-iron sheep herder's stove looked as though a very large elephant had stepped on it, and the rustic table and bench the Judge had so carefully constructed were all kindling wood. Sorrow-fully the Judge reached in among the rags of his recently neatly pitched tent and picked up the fragments of what had been a beautiful, very expensive fly rod. But the crowning touch to the gloom was the sight of an almost full case of fine Scotch whisky smashed as flat as a pancake with only a rich aroma left to remind them of its sweet potency.

With a typical rebound of good spirits, Butcher took the Judge's arm and said, "I'm dashed sorry, old man. Rotten luck, what?" And then he added with a Shakespearean flourish, "Come, we must not stand here mourning. Let us go repair the damage and celebrate in royal fashion a friendship welded even stronger by the whims of fate!"

We all got into the car and drove to my house, where I got out to watch them go on toward town in a swirling cloud of dust. The "repairing and celebration" lasted for three days, and their trail took them about 180 miles north to Calgary. There Butcher bought the Judge an even better English fly rod, a new tent, and complete camp equipment of the very best quality, all of which was brought back and put into use at the site of the wreck.

One evening I rode past the Judge's camp on my way to get the cows. All traces of the wreckage had been removed and the place was again neat as a pin. The Judge invited me to get down and come into the tent. As he showed me the various items of his new equipment with obvious enjoyment, he was warm and friendly.

Finally he turned and drew himself up as though making some kind of dissertation in court, and his words poured out in a rich Scottish accent that still comes back to me forty-odd years later with clarity and poignancy.

"My boy," he said, "apart from the fact that I give you my belated thanks for possibly saving my life, albeit in a somewhat unusual and unexpected fashion, I welcome you as a friend. Let's you and I remember that clouds of adversity most generally melt into sunshine, given a bit of time, and that which may seem like disaster may well be the source of better things." And having got that statement of profound wisdom off his chest, he cleared his throat with a great harrumph, grinned like a boy at me, and invited me to share his supper.

Many times after that we met by the streams, and the Judge introduced me to the art of fly fishing for trout. His was the first real fly rod I ever held in my hand, and under his direction I learned how to let the smooth silk of a fine-braided fly line go shooting on its own weight through the guides, so the fly came down lightly on the water with all the guile necessary for enticing the most wary fish. Thus he opened a door revealing something of worth and enjoyment in living—a door still open, which has led me to much adventure and exploration.

ALTHOUGH IT WAS SOME TIME before I contrived to get a fly rod of my own, its ownership was a means of continuation and enjoyment of a wonderful kind of sport. Like many of the oldtimers in the country then, a large percentage of fishermen today still feel that fly fishing is either a somewhat inadequate and impractical sport, or one that is too hard to learn; so they never try it. Actually, with about half an hour of rudimentary instruction, almost anyone can learn to cast a fly well enough to catch trout. In good time a person can proceed toward refinements that make the game more enjoyable. Apart from basic necessities, such as a proper rod, reel, and line, all one needs is four or five different patterns of flies in two or three sizes, although few dyed-in-the-wool fly fishermen will admit it. However, sooner or later, most of us use dozens of patterns; and if we have progressed to the point of hand crafting our own flies, the collection of materials used can represent a fair cross section of the fauna of the world. Fly fishing can be both an art and a science combined, for it entails not only a considerable study of insect life but also delicate artistry in copying various kinds in fur, silk, and feathers. It is a game that can be as simple or as complicated as anyone wants to make it.

Beginning with that first monstrosity tied with sewing thread, wool, and chicken feathers, my adventures among mountain waters took me into a phase of sport I had not known existed—a delightful combination of the development of equipment, art, and sheer love for the game. Through it I met many people from many corners of the world—the kind who enrich one's enjoyment of living by mere contact with them, the breed one enjoys trading talk with around the flickering light of a campfire. At the same time I encountered fish—individual trout—that time cannot erase from memory. In perfect recall focus are bright mornings when the snow-streaked peaks stood out sharp and clear against the vivid blue bowl of the sky, birds sang in the nearby groves of wilderness timber, and the air was tangy with the smell of pine and spruce and silver willow.

I remember one of those perfect days when everything seemed just right. My creel hung comfortably heavy on my shoulder after two full hours of fast action on the creek. The sunlit hills rolling gently to the foot of the Rockies were a mass of soft greens mingled with moving cloud shadows, and the trees sang softly in the breeze to the deeper accompanying overtones of the mountain stream. I was about ready to head for home. More out of habit than anything else I slipped quietly down through a grove of tall aspens and cottonwoods to a special place where a big pool lay against a timbered low-lying bank on a bend for another try at a mighty trout that made this place his home.

This was a giant fish—a rainbow trout—a real old tackle-busting he-trout, with a hook in his lower jaw like a salmon and a temper like a wild stallion. Over the course of three seasons, he had taken my fly several times and then broken off. My luck was bad with this fish, mostly because it was nearly impossible to keep him out of the network of roots under the overhanging bank where he had his lair. Fish for him with tackle heavy enough to hold him and he would ignore the lure as though it didn't exist. Work the kind of leader it took to fool him and he broke it every time. If trout could hold degrees, this one had his Ph.D. in survival. He was more than simply smart: he was uncanny.

The first time I saw him was when a puff of wind lifted my back cast high and hung my last remaining fanwing Royal Coachman in a tree twenty feet from the ground. That fly had been doing well for me and I was loath to break it off; so leaving my rod against a rock, I scrambled up the cottonwood.

After loosening the fly and letting it drop onto the gravel bar, I looked down over the pool spread out as clear as glass at an angle where surface reflections were nil and every detail of the bottom was in sharp focus. A half dozen fat cutthroats hung a foot or so above bottom ahead of a school of thirty or forty whitefish. There was the small dimple of a rise where the slow current eddied around the foot of a big cottonwood beyond. For a moment nothing was visible in the black shadow beneath the slanting tree trunk, but then a big, dark-green shape took form in a patch of sunlight.

At first I couldn't believe it was a trout—it was just too big; but then the thing moved—and an enormous rainbow

drifted lazily to the surface to suck in another floating natural; I almost fell out of the tree.

Maybe I scared him down with a sour cast or perhaps the big trout just quit feeding, for in spite of working on him till almost dark, I could not get as much as a curious look out of that fish.

A few days later I came back and floated a Brown Bi-visible over his hide. This time Lady Luck smiled a little, for the rainbow rose and engulfed my fly with the confident movement of a big trout that is completely fooled. But the smile was brief, for he fell on my leader in his first wild jump, breaking it like so much cobweb. Those were the days when we used diamond-drawn silkworm gut for leaders, with a relatively low tensile strength. In waters like this the odds were very much in favor of such a fish.

That was the last of him for that season, but the following summer he was back under the tree again, and I renewed acquaintance by sinking the barb of a tiny Grey Adams in his jaw. This time I stayed with him until the second round, when he dove behind a root, leaving me standing up to my knees in the riffle shaking like a leaf while I tied a new leader on the end of my line.

Another day I stalked him during a shower, when big drops were pounding the surface of the pool after a long, hot, sultry afternoon. The sudden change of temperature seemed to waken the trout, for they began to feed in a frenzy of action. Twice as I worked to tie a new fluffy Bi-visible on my tippet, the shattering rise of the big rainbow threw water in the air. His broad back and dorsal fin came right out and once I got a good look at a tail as broad as my hand. I was having trouble with my knot, for the sight of him brought on a mild case of buck fever. Keeping the fly dry was utterly impossible, but I hoped it wouldn't matter.

Shooting it upstream, I let it come down half drowned toward the trout. It swirled in a bit of eddy and the giant trout came up with his mouth wide open to suck it in. Like a tyro, I struck too soon and too hard, and once again the battle was his before it even started.

On and off all winter I saw that big rainbow in my dreams, and by the time the water cleared after the spring flood I was back armed with my experience and some specially selected leaders that I had tied and tested for just this purpose.

Now I was cool and bound to take plenty of time before making any move. Coming to the edge of the trees across the pool from his hide I stood among knee-high helabore and cow cabbage watching a hatch of drakes dancing on the surface. Wondering if the big trout was back from wherever he went in winter, I watched several small trout rising steadily upstream from where I was standing, then swung my attention to the foot of the big tree across the pool. As though in welcome, a big shadow detached itself from surrounding gloom to come up through the gin-clear water for a bug.

Then I heard the click of a reel coming from the tail of the pool: the place was already taken by another fisherman. Standing a few yards below where the pool broke over the edge of a fast riffle, a stranger was fast to a lively cutthroat with his light fly rod dancing in a graceful curve. He seemed as unaware of the big trout as he was of me. I stood motionless to watch.

The cutthroat came reluctantly to net and after killing and creeling it, the stranger dried his fly with several false casts as he made a step or two upstream. Again the shadow moved under the tree, and again there was little to give away the size of the fish, for there was only a tiny dimple of disturbance on the surface and a sound like a bubble bursting. But the fisherman saw it, and with flawless form shot his fly a little above and a shade to one side of the rainbow's lair. Holding my breath, I saw it settle like dandelion fluff and dance jauntily down the current without the slightest drag. Just short of the tree the fly spun through a couple of tiny swirls in a fairy dance, and then it vanished as the water boiled in response to the stranger's strike.

Like a short-grass bronc with a burr under the saddle and a tin can tied to his tail, the trout came lancing up into the sun amid a shower of water drops in a wild end-swapping leap that almost put him high and dry on a litter of driftwood along the bank. Somehow the leader held as the snappy little rod turned the big fish to the middle of the pool, where he jumped again—once—twice—three times in quick succession. Each time he jumped I saw the stranger throw slack to him, and I knew I was watching a master; for that is a trick known to few Rocky Mountain anglers and practiced by fewer still. The hardest thing I ever learned to do was deliberately throw slack to a big rainbow jumping on a short line. It's a crucial move, for it greatly reduces the chance of a broken leader or a thrown fly.

The rainbow had more tricks in his bag, which he brought out with sizzling speed. Time and again he made stampeding rushes up the pool, while the angler used the rod and current to wear him down. Several times he bored deeply toward his hide, but the gallant little fly rod bent to the grip as it turned him back. Once he jumped right at the man's knees, but when the spray settled the rod still danced in a fighting curve.

Finally the trout's rushes slowed to a weary slugging, which was still dangerous but without the smash and drive of his first efforts. I felt like cheering—maybe a little like crying too—as the man worked him slowly down toward the net. Grudgingly the big trout rolled on his side and slid head first into the open meshes. The net was too small for him, but it didn't matter; the fish was completely exhausted —too tired to even wiggle.

I was reaching for my camera in my jacket pocket, when I saw something beyond belief. The man downstream carefully slid the rainbow from the net, held him up for brief admiration, unhooked him gently, and, while the trout worked its gills gathering a reviving supply of oxygen, stroked him as he held him upright in the current. Then with a certain majesty and a slow swirl of its tail, the big trout headed back for its hole under the bank.

"Well I'll be damned!" I said to the world at large.

It was the stranger's turn to be surprised. "Where did you come from?" he asked.

"Right here for the whole show," I told him. "Why did you turn him loose?"

The man smiled and replied, "I've been taking a crack at that big bruiser every time I get the chance for quite a while. Hooked him several times but he always beat me. Killing him seemed like a kind of fool way of celebrating. Anyway, it's fun to know of a place where there is a really big fish."

"You're right," I said, as I went over to introduce myself and shake his hand. "That was some fight!"

Several guiding trips with the packtrain kept me away for the rest of that summer and fall, and no chance came to go back that season. I never saw the big trout again. Maybe someone caught him. Or perhaps a mink killed him. It could be that he changed his living place for another pool, where he finally died of old age. Whatever happened, that big trout taught me much, and if the man who caught him and turned him loose should see this book, I would like him to know that he also showed me something of value. It is possible to have your cake and eat it too; but sometimes it is much better to forego the latter pleasure for the sake of keeping alive something of great beauty for a while longer.


OVER THE YEARS my trails have led me through some of the wildest and most rugged reaches of the mountains all the way to the tundra prairies of the high Arctic. I have fished British Columbia's Bella Coola in April near spires of peaks hung with snow and clear, blue ice, when its bottom was literally paved in places with big fighting steelhead—their flanks like molten silver—fresh from the sea. I have known the indescribable feeling of one of these fish tearing the river wide open as it plunged and ran in its wild bid for freedom.

I have cast a tiny dry fly over northern waters so stiff with pastel-shaded arctic grayling that it was merely a question of which fish got to it first. The kind of fly couldn't have mattered less; the size of the fish one hooked depended on its proximity to the lure and its speed.

Once a small ten incher was fighting in fast water at the foot of a falls when up out of the depths came a great char—a lake trout that had taken to the river—and engulfed the grayling, the tiny fly, and a good portion of the leader. My rod weighed a mere two and a quarter ounces, and the resulting battle threatened to smash it a dozen times, but the eight-pound fish finally rolled over in defeat.

No matter the thrill, when the campfires flicker at night and the little red gods dancing in the flames encourage recall of the past, I see trout coming to crude lures presented by boys lucky enough to have been in places where boys should grow up.

# Along Trapline Trails

Snow on the level, three feet deep,
Oh, Lord! How the wind is blowing!
We've eaten our caps and suspender straps,
And damned if it still ain't snowing.

We made a mulligan stew today,
Of a candle, some soap, and a wisp of hay,
Some small pine blocks, a pair of socks,
A wood rat's nest and a couple of rocks.

*Tomorrow morn at the peep of dawn*
*We're going to leave this shack,*
*A trail to seek o'er mountain peak,*
*Adios! If we don't come back.*

*And in the spring when all nature sings,*
*Should you chance on this trail to come back,*
*If you find a couple of skeleton things,*
*You'll know it's just me and Jack.*

YEARS AGO, CONRAD KAIN, ONE OF THE MOST FAMOUS OF ALL mountain guides, found this poem written on a scrap of paper in a long abandoned trapper's cabin, while on an exploration trip in the Canadian Rockies. It is flavored with humor and a strong touch of toughness—characteristic of the breed of trappers—and it leaves one wondering if the two hungry men were successful in their break over the mountains for some grub.

In the sketchy history left behind, which is fascinating and at the same time tantalizing in its almost inevitable brevity, the era of the trapper is colored with daring, adventure, vision, imagination, and just plain guts. The trapper was the one who braved the storms, the hostile Indians, and unknown country to blaze those first trails over mountain passes later used by migrating settlers in their quest of new opportunities and new land in Oregon, Washington, and California. He rode horses into the far reaches of the mountains and to the beaches of the Pacific Ocean in his quest for beaver. He fought his way on foot through the devil's club thickets and quaking muskegs among the ranges west of the Continental Divide along the headwaters of the Peace, Columbia, and Fraser Rivers. He paddled his canoe down brawling unknown rivers clear to the Arctic Ocean. It was he who established the beaver trade—from which fortunes were made for the Hudson's Bay Company and the Astors. And in the process of their quests many of the trappers died with their moccasins on—their whitening bones scattered by foxes, bears, and other animals. If they had been given a choice this was the way they would have wanted it, for they were about as wild as the animals they pursued; most of them reveled in the freedom and action their lives dished up to them; and very few worried a moment about their trails concluding in an unmarked grave or no grave at all.

Once years ago, while following my trapline, I found an ancient, hand-forged steel trap half grown into a tall aspen tree, testimony that another had lifted the skins of animals along the same creek perhaps half a century before. Forty annual rings of growth were in the wood enclosing the trap, and one could only guess how long the trap had hung there before the tree began to enclose it.

There is still a scattering of this hardy breed of men strung out through the wilds of northern Canada from the rocky shores of Labrador to the coast of British Columbia. No longer are fortunes made by gathering the silky skins of fisher, marten, fox, mink, and beaver, for the introduction of synthetics and the vagaries of the trade have spelled sunset to the golden era of the trapper. But there are still men who cling to this way of life, enduring the blizzards, hunger, and hardship—the price paid for the freedom and independence of wilderness trails.

They know few comforts. They ask little of life except the opportunity to establish a territory with a lot of living space in some wild valley with perhaps only the jays for company. They are tough enough to stay alive in spite of sometimes appalling hardships, and they need only enough money to trade for a grubstake with maybe something left over for some gifts for a woman, or whisky for the makings of a celebration.

Bert Riggall, while on an exploration trip by packtrain through the mountain country between Fort Steele and Pine Pass in British Columbia in 1911, recorded an interesting observation. Somewhere in the wild muskeg country northwest of Yellowhead Pass in central eastern British Columbia, he and his partner Cyril Watmough caught up to an old trapper. He was relaying his winter supplies from the railroad to his base cabin by backpack on foot across some of the roughest country to be found on the continent, a matter of many miles and much weight, so the job was due to use up most of the summer and fall.

They had several horses traveling light and one or two carrying empty pack saddles, so they loaded up the old trapper's grub and gear, thus trading him transportation for his services as a guide through a piece of country that was heavily timbered, strewn with down timber and boggy in most of the open places. It was bad going for a man on foot and about the limit for horses. Even so, they reached his cabin in only a few days, saving him a great deal of hard travel and back-breaking work.

The old trapper was very grateful, for the years of hardship and poor food were taking their toll on his aging frame. He was suffering from rheumatism and stomach ulcers. Bert mentioned seeing two or three of his line cabins, which were not much better than dens—just trenches dug into a bank or slope covered with A-shaped roofs made of poles. He didn't use a stove, the cooking and heating fire for overnight stops being built in a shallow hole on the floor with the smoke issuing through another hole in the roof. A crude bunk served for a place to lie down and sleep.

With any amount of building logs available for the cutting, such crude and comfortless lodgings seemed unreasonable. When asked why he didn't build more comfortable cabins, the old man replied, "If you are going to make any money at this game, you have to get out and work in all kinds of weather. If a cabin is too snug and warm, a man is liable to hole up in a storm instead of getting out and tending his traps. A man wants to have his cabins just a little worse than outside; then he gets out and works!"

Such a philosophy had served the old man for a long time but may have finally spelled his doom, for nobody ever saw him again. Somewhere in the rotting wreckage of one of his line camps or perhaps moldering among spruce needles on the forest floor, his bones remain hidden to all but the birds and the squirrels.

Approach to environment and the ability to live comfortably within its wide variations differ widely among individuals. Some men, like Joe Cosley, lived easily and with some comfort in country that promised nothing but the most abject hardship during the hard months of winter. Others with years of experience and equal opportunities to learn how to take care of themselves merely endured almost killing conditions without any apparent effort to make their way of going easier. Some of these no doubt took pride in their toughness, but many just failed to learn how to cope with the wilderness.

Years ago, when I first began rambling with the pack-train during the summer and fall through the wilds of the upper Flathead River drainage in British Columbia, I found places where someone had left a charred log seven or eight feet long split in two pieces. There was nothing else—just a split charred log, and for a time I was baffled. But I suspected they were a part of the winter activities of Levi Ashman, who trapped the region. When he came to work for me as guide, I asked him about the charred logs.

He explained that these marked the sites of overnight camps made on the trail in winter, when darkness caught him out of reach of one of his cabins. He simply flattened a place in the deep snow with his snowshoes in some fairly well-sheltered spot, and then built a fire in the middle of it. While a kettle was melting snow for his inevitable tea, he would cut and split a section of log—generally a piece of green pine about seven or eight inches in diameter. While he prepared and ate his supper, which usually consisted of tea and moose meat, he sat on one log and used the other half for a sort of crude table. By the time his supper was cooked and eaten, the fire had burned down into a bed of coals at the bottom of a hole melted out of the snow lying several feet deep. While he worked on the skins of his day's catch, Levi kept adding wood now and then. By the time he was ready for some sleep, the hole was deeper with a bed of coals glowing red in its bottom. Then Levi simply placed his split logs side by side, flat side up over the hole, lay down on them, and covered himself with a piece of canvas, for he carried no sleeping bag or blankets. He did not even bother to remove his leather-topped rubbers, but slept fully dressed. On very cold nights he might have to get up several times to add fuel to his fire. Regardless of temperatures he was well smoked upon hitting the trail in the morning.

"It's a wonder you haven't burned yourself up," I remarked.

Levi gave a dry chuckle. "Burned the seat out of my pants one time when it was about thirty below. Damn near froze my rump before I made it back to my cabin next day. Broke a snowshoe and the going was mighty slow. Reckon the extra work kept me from freezing."

Levi was a remarkable character in more ways than one. We were moving the pack outfit through his trapping territory at the tail end of the hunting season one fall when I spotted him butchering a young bull moose on a gravel bar across the Kishaneena Creek. Letting the crew take the pack-train ahead, I rode across to see if there was anything I could do to help him. There he was, covered with blood and hair, up to his elbows in moose innards, busier than a beaver at freeze-up time. My horse snorted nervously and my nose told me something was wrong. It became immediately obvious that the moose had been dead for some time before Levi had begun dressing it. When I remarked casually on this, Levi explained that he had killed the moose the previous evening. Then he had chopped down a tree to make a bridge across the river so he could dress it. But he chose a springy green pine instead of a more stable dry one, which was a mistake.

"When I was right in the middle of it," Levi told me, "the durn thing got to jumping and bucked me off clear over my head in the water!"

By the time he got back to his cabin, dried out, and had something to eat, it was too dark to go back for the moose, so he just left it till next day. Assuring me that it might taste a bit gamey, but that it would be tender, Levi kept right on with his butchering job.

When he got it chopped into pieces, I hitched my lariat to the load of meat to drag it the half mile to his cabin. Levi refused to ride double with me across the stream, not trusting my horse, which was making a fuss over having to drag the moose meat. When I was about halfway across the river, I looked around just in time to see Levi once again sitting in the ice cold water up to his ears under his bridge. Swearing and chuckling in a sort of accompaniment to his chattering teeth, just as though he had pulled the whole thing off for my special entertainment, he came trotting through the snow behind my horse up the trail, his small spare frame looking deceptively frail in his wet, clinging clothes.

I never did find out for sure if he ate that moose meat or used it for trapline bait and killed another for his winter's meat. Furthermore my curiosity did not run to the length of dropping around for a visit to find out. Levi was in many ways one of the gentlest, most good-natured men I have ever known, but he was also rawhide tough and thoroughly capable of enduring most anything, including high moose meat.

THE EARLIEST TRAPPERS who operated on the eastern edge of the Rockies and along the rivers flowing across the adjacent plains to the east lived for the most part on fat buffalo, elk, and beaver. Their main concern was keeping their hair from becoming the lodge decorations of some enterprising warrior. When the beaver were gone from the more easily accessible places, the trappers spread out into the mountains and northward to the frozen prairies of the Arctic. It was there that the elements became the prime force to be reckoned with, and many the tale of hardship and endurance has been told when trappers ran low on luck or made some wrong guesses.

At best trapping is a young man's game, although there are men who stuck to it till the sands of time ran out for them. It is not all hardship, for, like many things, hardship is relative and what may be the toughest going for one man may be nothing but a lark for another. It has a certain lure of adventure that attracts the inveterate prospector to keep going, an aura of the unknown and a gamble as to what one may find around the next bend of the trail. If a man is lucky enough to find a woman who is willing to share his cabin and the adventure, he is indeed fortunate. Furthermore, the environment is pure heaven for children trained to take care of themselves in the wilderness. A man who has lived thus knows the full satisfaction of the primitive—the underlying characteristic latent in all of us—when he comes in off a long trail with a fine catch to find a warm cabin, supper on the stove, and a serene, happy woman to exclaim over the rich, silky furs.

Nothing quite matches hitting the trail on a frosty morning when the low sun is just painting the tops of the hills and mountains with pale rose; and the only sound to be heard comes from the hollows where the somber light of dawn still shrouds the timber—the sharp intermittent snap of frozen trees. One is aware of the creaking of the webs and the soft swish of snowshoes in the two or three inches of new snow over the two or more feet of winter accumulation. It is the perfect combination for making fast time over rough country through and between the hills, across the frozen swamps and snow-draped forest glades.

Far from being an impediment, the snow makes the going easier, as it covers logs, fills holes, and smooths out the country for travel on the webs. It also acts as a fresh clean page after each snowfall, on which is written in minute detail the track stories of the activities of every living thing moving across it. Here and there close to dead logs, or where the wind has whipped the snow away to reveal projecting tufts of grass, is the lacy filigree of mouse tracks, where these small animals came up through the snow in search of feed. There are the deep indents of a cow moose and her calf, or perhaps those of deer or elk, in their erratic wandering in search of browse. The broad furry pad marks of snowshoe rabbits interlace the willow thickets. Through all these patterns are the interwoven tracks of the hunters: coyotes, foxes, maybe wolves and lynx. There are also those of the marten, weasel, mink, fisher, and squirrel. The trapper, who is also a hunter living off hunters, notes all these signs and if they are numerous he is a light-hearted traveler, there being a certain satisfaction to moving through fur-rich country.

Like all youngsters, my brother and I had a need on occasion for spending money. When I was seven or eight years old I had a sudden and burning need for a toboggan. It cost all of five dollars, a fortune for a small boy—more money than I had ever had at one time. It posed a problem of mountainous proportions, but this did not result in the desire being completely snowed under; rather, it whetted it. As fathers can do, mine became acquainted with my problem, thus sharing it, for he too had once been a small boy and had not forgotten what it felt like. He pondered the problem gravely, but instead of giving me the money for the toboggan, he took about half the sum and bought me some small traps. He showed me how to set these for weasels and mink and thus I became a trapper, running my own business and knowing the satisfaction of being financially solvent most of the time.

My first trapline ran through the willow and aspen groves between home and school, and quite often upon arriving for classes there was a weasel in my bag along with books and lunch. Before long some of the other boys were also running traplines on the way to school, and sometimes we smelled a bit gamey from contact with the animals we caught.

Some teachers frowned on the practice, but there was not a great deal they could do about it. One young lady who taught us got a gleam in her eye when she saw some of the snow-white weasels with their black-tipped tails. With an ermine stole in mind and some lessons from her pupils, she set her own trapline and began to collect the makings of an exotic fur piece.

All went well until one morning, when she failed to show up at the usual time. It was a cold morning and we were all inside the one-room building, with the heater going full blast, when she came staggering in the door, pale, glassy-eyed, and obviously fighting back an inclination to collapse. We stood looking at her in astonishment wondering what the matter was, when all of a sudden the smell of skunk almost flattened us. Somehow we had forgotten to tell her what to do in the event she caught one of these highly potent animals, and she had tried to kill it with a stick. Weakly she managed to gasp that school was dismissed and then she climbed back on her horse and headed for home. We got a couple of extra holidays that month. It was the only time I recall a skunk being so benevolent.

Sometime in April when the ice thawed on the sloughs, we set our traps for muskrats, which teemed in the little lakes that dotted the country. It was amazing how much fur small boys could collect in two or three weeks working in the evenings after school and on weekends. It was a time when ranch mothers put up with considerable extra work; for inevitably we got muddy and soaked on numerous occasions while trapping 'rats. But nobody complained, for furs sold for good hard cash, and no matter who was making it, the extra money helped out. Skinning our catches was always a messy business, but the more we practiced the better a job we did, until we could shuck a muskrat out of his pelt in no time and pull it over a wooden stretcher like a damp sock to be cleaned of excess flesh and fat, and then dried.

Besides being profitable, there was freedom to be enjoyed while rambling a trapline. So I went trapping for a living when I quit high school in the midst of the great depression. It was a time when ranch hands were working for their board and tobacco. Grass-fat steers weighing a thousand pounds were selling at ten dollars a head. Vast numbers of people in the cities and towns were on government relief, and nobody was able to get hold of very much hard cash.

But, strangely enough in the midst of this financial drought, raw furs were bringing a good price. What was even stranger, very few people were taking advantage of it. The country was wide open for an enterprising trapper. All one needed was a grubstake and some traps. These took some money.

Late in August I hitched up a good gentle team to a wagon and hayrack, and with my bedroll and lots of enthusiasm headed down out of the hills to the grain country on the plains looking for a job with a threshing outfit. I was lucky, for I did not have to look very far before getting on a big outfit at five dollars a day.

There were twelve teamsters working on that rig. The weather was abnormally hot and the crops heavy. With wheat running at sixty bushels to an acre and oats between eighty and a hundred, the bundles weighed heavy first thing in the morning and like lead by night. We were going steadily seven days a week from starlight to starlight. I was only a stringy kid of sixteen, and for three days I thought I would probably die. Years later, the boss of that outfit told me the rest of the crew were laying bets as to how long I would last and the most optimistic of them gave me two days. Likely several people came very close to collecting their wagers, but somehow I kept going.

Then like magic things began to get easier as I got the hang of using a bundle fork and keeping up without wasting any effort. From what amounted to pure torture, the job became just hard work. We moved from farm to farm eating prodigiously off the cream of the land. We had the chance to find out which of the ladies were the best cooks. There were no really bad ones—some were just better than others—and to the last one they tried to outdo each other in quantity. Most of them—even one of their daughters on occasion—had a smile for the kid who was hollow clear to the knees. I never seemed to get filled up, for no matter if I ate till I could not hold another crumb, an hour later my belly was grumbling for more food. We all ate prodigiously three times a day, plus a big lunch served in the field in midafternoon. Each night when I eased into my bedroll I could count five more dollars in the bank roll—no fortune by today's standards, whereby a boy can earn that much running errands for an hour. But at that period of history it was enough to buy a good pair of pants with some change coming in the deal; a box of .22 rifle ammunition cost thirty-five cents; one could purchase a good pair of work boots for five dollars; and a pair of Indian-made buckskin moccasins cost a dollar.

By the time the threshing season was over thirty days' wages were coming to me, enough to buy part of what was needed, so the rest of my outfit was bought on credit.

That was the first of many winters spent following trapline trails among the foothills of the Rockies. Traps were used to take mink, weasels, and muskrats along with the occasional coyote and fox. Most of the coyote pelts were gathered with wire snares and the rifle. The snares were set along trails used by the little gray wolves, one end fastened to a tree or drag, and the other adjusted in a loop to go over the coyote's head. They were easy to set and much more effective than a steel trap in the fast-changing temperatures of the chinook belt, where it could be twenty below in the morning and sixty degrees warmer by supper-time. The bad thing about snares was their complete non-selectivity and their strength. They would hold about any-thing that got hung up in them. Once I found a deer hanging dead in one, and another time I caught a wild old cow by the leg and had a real battle turning her loose. To be effective a coyote trap had to be concealed in dry earth or snow. If the cover stayed dry, it continued to be effective, but if it got wet and then froze in a sudden change of temperature, it could be so iron hard that an elephant could step on the trap without setting it off.

OF ALL THE ANIMALS a trapper encounters, the coyote is by far the most cunning and intelligent. I have trapped foxes, which have a reputation for being difficult to take, but com-pared to a coyote the fox is a dunce. The wolverine has been credited with being almost supernatural in its ability to dodge traps. The ones that are the most difficult to take are those that have been conditioned by getting pinched in small traps while stealing the bait from sets made for martin, mink, and such small animals. Ordinarily a wolverine is no problem to a good trapper. The trapper who can make a set with some confidence that the next coyote that comes along is as good as caught is a master.

Ever since the white man came west he has been waging war on the coyote with guns, traps, snares, dogs, and the most potent of poisons; but the little gray wolf is still found over large portions of its original range, and he is learning to take better care of himself every day.

A coyote can detect a carelessly handled, dirty trap under a half inch of dry sand no matter how well it is hidden. His eyes and ears are good and his nose superlative. To trap coyotes with any hope of success, the traps must be boiled for at least two hours in a strong solution of wood ashes, scrubbed free of rust and dirt with a wire brush, rinsed in clean water and then reboiled in a soup made of the bark of trees, shrubs, and grass found in the area of the trapline. This treatment of blending the smells of surrounding herbage with the steel makes the trap very hard to detect. A coyote trapper never handles his traps with bare hands, but always with clean cotton or buckskin gloves.

Even with all these precautions, it is easy to give away one's strategy. Once during a blizzard I made what appeared to be a completely foolproof set for a coyote, which turned out to be something else. Taking three clean traps, I wired them together by their chain rings and then set them. Riding a gentle horse and carefully carrying the traps, I leaned down from the saddle and planted them in the snow beside a horse carcass that was within sight of the cabin. The blowing snow was a perfect camouflage.

Next morning at daylight I looked out just in time to see a coyote coming toward the bait. Stopping frequently to look all around, the little gray wolf carefully approached his breakfast from the downwind side. When still about one hundred yards away, he stopped, gave the place a good going over with eyes and nose, and then circled the dead horse completely. Arriving back at his original line of approach, he once more approached with caution—step by stealthy step to within two feet of the traps. There he suddenly froze with forepaw uplifted like a pointer dog on birds, his nose going slowly down toward the snow while his tail came up to half mast. Through my binoculars I saw his back ruff lift into a standing roach of fur, and then he began to back out of there, placing his feet with the utmost precision in the incoming tracks until he was well clear. Then he swapped ends and lit out as though the devil was on his tail.

Wondering what had gone wrong, I rode my horse down for a closer look. At first I was baffled, for everything looked perfect, but a closer inspection revealed the danger signal. Barely visible, sticking up out of the windswept snow, was the stem of a match that had fallen from my shirt pocket when I had leaned down to place the traps. It was a minute detail, but it likely smelled of me, which made it about as conspicuous as a telephone pole to the coyote! Such carelessness costs a trapper money, for it educates the individual coyote concerned to a real genius among little gray wolves.

Further illustrating the adaptability and intelligence of the coyote, I observed a wilderness incident one cold winter afternoon that was about the ultimate. It was about twenty-five degrees below zero, with a foot of snow on the ground and a northeast wind blowing. I was hunkered down on top of a little hill among some windblown scrub pines watching a big bull elk bedded down on a bench on the mountain face. He had a mighty handsome set of antlers and enough meat on the hoof to keep my family going for months; but he was safe from my rifle, since he was lying just inside the boundary fence of Waterton Lakes National Park.

Slowly but surely solidifying, I endured the icy breeze in hopes that the bull would get up and feed down into legal shooting ground before it got too dark to see the rifle sights. The sun had dropped behind the cold, gray, cathedral shape of a peak in the background, and the frost crystals were flowing in the wind like smoke haze in the air. It had been cold all day, but now it was fast growing really bitter. About the time I was ready to give up, the bull stood up in his bed to stand gazing down—a regal animal with his great antlers sharply etched in silhouette against the snow. Then he began to feed, but instead of coming down he stayed on the grassy bench just under the spot where he had been bedded, pawing down through the snow to the rich bunch grass underneath.

While I watched, a coyote came out of the thick scrub farther up the mountain to stand studying the bull elk. Suddenly he seemed to come to a decision, for he came trotting down toward the elk in a businesslike fashion with all the confidence of someone who knows exactly what the demands of the situation required. What he figured it called for was a furry, four-legged mouse eater with a handsome brush and sharp ears, for, after giving the bull a close look and being totally ignored, he took up a position just a bit to the rear and close alongside the big animal's flank. From there he proceeded to watch for mice stirred up by the elk's pawing for grass. Twice I saw him leap almost completely under the bull's belly after a mouse, and he was apparently successful in catching the small animals. It was incongruous to see that great proud stag acting as a sort of unaware mouse digger for the enterprising coyote. Quite obviously this was nothing new for the coyote. He had caught mice this way before—perhaps even with the same bull.

It was almost dark, and obviously my hunt was at an end, so I slung my rifle over my shoulder and pointed my snowshoes for home, empty-handed after a long cold day on the trail, but warmed by having seen something of interest.

Generally speaking, coyotes are very smart animals, but certain individuals sometimes develop a degree of intelligence far above anything normally attributed to their kind. Contact with man is nearly always directly responsible for this development.

One winter I caught a big female coyote in a trap set for

a fox, and she escaped by twisting off a toe from a front foot. Her track thus became recognizable, but she developed into an expert at avoiding traps. She offered an irresistible challenge, and for two seasons she and I matched wits in a one-sided battle wherein she won every skirmish.

I tried every kind of set I ever heard of and some that were my own invention without ever really coming close to tangling her up in steel. To make her even more aware of my intentions, I shot at her one morning and missed. From then on about all I ever saw of her was her tracks. What a wonderfully cunning animal she became! Knowing this did not lessen my covetous plans to take her big silky gray skin from her back.

One cold day when there was a new skiff of snow on the ground, I was crossing the ice of a lake along my line when I spotted her telltale tracks. She had been investigating the muskrat push-ups dotting the lake, going from one to another sniffing the muskrats inside their frozen walls and apparently enjoying herself without much chance of catching anything to eat. Since I knew that she made a patrol of this part of the country almost nightly, the sight of her sign and her interest in muskrats gave me an idea.

Push-ups are the feeding stations of these aquatic animals. Muskrats live in houses something like smaller versions of beaver lodges, and, like beaver, they do not make a habit of feeding there. When the ice first freezes in late fall and is still thin and rubbery, they push wads of aquatic weeds from underneath up onto the frozen surface until a hollow mound of this material is constructed; it freezes into a hard shell and eventually dries out on the outside surface, thus creating a layer of insulation to keep it from falling apart during warm spells. At the same time the open hole under it is kept from freezing, so the muskrats have a cozy feeding station. If something threatens from above, the 'rats merely dive and swim away to safety. Even if the predator tears the top off the push-up, the muskrats usually get back before their plunge hole freezes solid and repair it quickly.

Going to one of the push-ups, I cut a small hole in the top of it with my belt ax, and set a small trap in it. Plugging the hole again so light would not shine through, I waited, and in less than half an hour had a muskrat in the trap. Tying a bit of rag to a stick, where it would frighten any other rats that showed up, I left the top of the push-up open and headed for home with the live muskrat in my pack sack. Upon returning with a suitable coyote trap some time later, I put a couple of big handfuls of dry grass into the push-up on top of the layer of hard ice that had frozen over the plunge hole in my absence. Then the live muskrat was turned loose inside the push-up and the top carefully fitted into place so he could not take off overland. For the time being the little animal was my prisoner and also a lure for a very cagey coyote. As I carefully buried a number 3 trap on top of the push-up I could hear the muskrat moving around inside. In the meantime the wind had whipped the snow off the lake ice, aiding me by removing all tracks.

When I returned the following day the trap was gone along with the stone wired to the chain for a drag. Skirting a strip of heavy willow growth along the lakeshore opposite my set I found a drag mark heading off the ice through soft snow. Not far beyond, the coyote was tangled up in a clump of brush. Sometime during the night she had located the set and had put her foot in my trap when she tried to get at the muskrat inside.

Taking her skin somehow lost its attraction, and when it was hung up with the rest of my collection of furs, I regretted killing her. For in the process of my long contest with this wise animal, she had taken on a certain character—an interesting part of the country along my trapline. Now the challenge was gone. No longer did her three-toed track show up. Something of value was missing, which no profit accrued from a pelt could replace. It was at such a time that I took a long hard look at my way of going, and somehow the looking left a bad taste in my mouth. It is probably a mistake for a trapper to get too close to the animals on which he preys.

To be a successful trapper one has to study his quarry, and so he becomes something of a naturalist, much more so than a hunter using firearms. The hunter's association with animals is not nearly so close; it is much more indirect and leaves him with less opportunity to observe. The trapper lives with animals in their choice of environment. The more he sees of them, the more he becomes aware of their intelligence, their warm-blooded ways of living, and their sometimes amazing ability to adapt on the instant to unusual circumstances.

One time I had a coyote trap set on an ant hill with a piece of rabbit and a flapping rag of its hide tied on an overhanging willow bush about five feet above it. Such a set will also take a fox, and if made close to a lake or stream even a mink on occasion. One morning after a fresh snowfall of about four inches a great horned owl was sitting solemnly on the ant heap with one foot hung up in this trap. When I approached the big bird fluffed up its feathers till it was twice as big as normal, hissing and snapping its beak fiercely. It was not much hurt, so I undertook to take it out of the trap alive, for I am fond of owls—especially great horned owls, magnificent winged hunters of the forests and extremely attractive. In handling an owl, one does not worry much about its beak, for they rarely bite in self-defense, but their long, curved, razor-sharp claws are something else. Given a chance they can bury these in a man's flesh in the wink of an eye, for they have enormous strength in their grip. They are a dangerous bird for a wilderness trapper to handle, for their inch-and-a-half long claws are inevitably fouled with fragments of meat from their kills, and any wound suffered from them is extremely susceptible to virulent infection. But by blindfolding this one with my jacket and using care, I was able to turn it loose. Resetting the trap, I headed on up the trail.

Only a few yards from the set, where the trail skirted through some open timber along the edge of the lake, I cut the fresh tracks of a coyote where it had been sitting watching the owl. It had moved on before my arrival, for its unhurried tracks headed the same direction as my trail out onto the open flank of a rugged ridge. Thinking I might collect it with my rifle, I trailed it up to where it went into a hollow. Beyond that I could see where it went over another rise of ground into a second dip. It was likely the coyote was still in this second hollow out of sight, for a look through my binoculars showed no tracks going over a little ridge rimming it on the far side.

Just as I let the glasses down in the carrying strap slung around my neck, I caught a glimpse of an eagle peeling off the edge of a white cloud five thousand feet up and come down in a sizzling stoop that accelerated into a crescendo of roaring pinions vibrating in the wind. It shot from sight into the far hollow and at the same instant the coyote appeared on the far ridge top, going down along it as fast as it could run toward the place where the eagle had disappeared from my view.

I ran toward the intervening ridge a hundred yards in front of me to see what was going on, but on cresting it there was nothing to be seen except the eagle flapping its way across the valley with empty claws. But the drama was written plainly in the snow. The eagle had come down to strike a big jackrabbit in its hide in a patch of prairie rose brush, but it was too heavy for the big bird to carry away. While the eagle was struggling with its catch, the coyote had come streaking in to hijack its kill. The coyote had then carried the rabbit into a dense strip of heavy brush along the bottom of the draw.

My rifle at the ready, I eased down to the edge of this cover and whistled. Almost instantly, the coyote came running out the far side going full speed, and at the skyline about two hundred yards away, it stopped to look back. This was a mistake, for my .250 Savage rifle cracked and the little gray wolf dropped where he stood.

Trailing him down into the willows I found the still warm body of the rabbit where it had been dropped. It was undamaged except for three great claw punctures through the loins at the kidneys. All in ten minutes, that rabbit had changed ownership three times, and an eagle had lost his lunch, a coyote his life.

OF ALL THE ACTIVITIES of trappers during the course of history, it is the pursuit of beaver that has offered the most color, adventure, and plain unadulterated hardship. This was not because the big furry rodents are particularly sly or difficult to catch. They were found in great numbers by the early trappers, and about the most difficult part of catching beavers has always been their environment and the hard work involved.

One can visualize a group of trappers—perhaps three or four—riding down into one of the river valleys from a pass between the high, snow-clad peaks of the Rockies in early spring. They are trailed by a number of pack horses carrying their "possibles": salt, tea, maybe some flour, powder and lead, jerked buffalo or deer meat, extra knives, traps, trinkets for presents for the Indians they might encounter, and other odds and ends. They lived with very little more than what the land could provide, these half-wild, buckskin-clad men, who stayed alive by hunting, trapping, and matching wits with the plains Indians. Picking a spot suitable for a camp, where there was shelter from sudden spring blizzards, plenty of grass for their horses, firewood, and some natural features of defense in case of attack, they set up camp. Camp could be a crude cabin, but was more likely an Indian lodge made from the tanned skins of two-year-old cow buffalo. Sometimes one or two of the party were accompanied by their Indian wives, who were very adept at setting up and keeping a lodge and experts at dressing skins.

It is impossible for a beaver trapper to set and attend beaver traps without getting into water, which is why modern trappers working in spring always wear hip rubbers. The old mountain trappers had no such comfortable footgear. Their buckskin breeches were so made that they could unlace their leggings at the knee and take off this lower covering. Then with their feet slightly protected from sharp sticks and stones by a pair of old moccasins, they just waded in and toughed out the freezing cold water.

In such virgin beaver country a trap set almost anywhere along a big stream like the Yellowstone or the Saskatchewan or smaller tributaries meant a beaver captured in a few hours. Catching all the beaver a man could skin, stretch, and flesh in a day was no trick, for it is the preparing of the skins that takes the time.

A beaver is reluctant to give up his skin even in death.

It is like skinning a bear or a badger, a matter of cutting with a sharp knife all the way, quarter inch by quarter inch, until the pelt is free. Experts with their big Green River knives, the trappers could shuck a beaver out of his hide with dispatch, but then it had to be laced into a willow hoop so that it was wrinkle-free and almost round. Any flesh or fat still sticking to the raw side of the pelt was then scraped off. It all took time. It was tedious, but nevertheless few trappers complained about this work, for it meant profit. When a group of men worked together, there were always one or two specialists with their knives who gladly traded their skinning skill for going without regular dips in the cold water. A durable trapper who did not mind cold water could keep a good skinner busy in top beaver country.

It was dangerous business in more ways than one, for apart from Indians, and grizzlies that might be attracted into camp by the smell of beaver, high water and avalanches threatened, and there were many ways a man could slip into an accident—a glancing knife or ax, an accidental plunge into fast water that ended up under a log jam, or a mixup with a horse. But in no way could these men be termed ac-cident-prone, for those that were never got as far as the beaver country along the foot of the Rockies.

While it cannot be said with complete truth that beaver can make water run uphill, they can be given full credit for trying. The beavers living in the larger streams do not build dams, but those that occupy the smaller streams do, and their industry will sometimes choke such a stream to the point where natural evaporation from the surface of the ponds will use up all the run-off. The actual tonnage of mud, sticks, and rocks they move and use in the construction of their dams, lodges, and network of tunnels and ditches is amazing. I have seen a dam close to three hundred yards long varying in height from one to twelve feet holding back a stream. The width of a dam at its base is about equal to its height or even a little more, and this method of construction makes the weight of the water wedge and hold the dam down. When one considers that a dam, lodge, and complex of tunnels and ditches are built cooperatively by a family group of from five to ten beavers varying in weight from twenty-five to seventy-five pounds each, with only their paws and teeth to work with, one begins to appreciate just how persistent, ingenious, and industrious these animals are.

Apart from the dam, they usually build a big dome-shaped lodge out of mud and chunks of wood by simply heaping it up in a place surrounded by deep water or on the bank fronting on a deep place, and then digging and chewing out its middle into a living chamber. Usually there is only one house in a pond, but occasionally two, and once I found a big duplex affair, measuring ninety feet over the dome from water line to water line, in which two families were living. The beavers also dig a system of canals and ditches as well as tunnels into the banks leading to feeding stations, for the house is strictly for living. A really big dam such as previously mentioned is not the work of a single season but of several.

Beavers are excellent swimmers and very difficult for a predator to handle in water. A pair weighing fifty or sixty pounds each with kits in the lodge have been known to drown a good-sized dog. Anyone allowing a retrieving dog to swim in a beaver pond during the spring months is putting his animal in grave danger. An attacking beaver simply comes up under him, grabs a mouthful of hide, and pulls the dog under.

My brother distinguished himself one day while fly fishing by hooking a beaver and then compounding the folly by having a nose to nose argument with a big one at a range of about eighteen inches.

It was a warm day in June. Everything was lush and green in new growth. We had the day to enjoy some fine sport fishing on a string of dams up near the head of Cotton-wood Creek. There were trout rising all over one big dam when we arrived, and while we were tying flies on our leaders, we were also aware that this was the home pond of a pair of large beavers. They took exception to our intrusion and the two of them kept swimming back and forth looking us over and alternately diving with resounding bangs of their tails on the surface. This sort of thing is not exactly compatible with fly fishing, for they were disturbing the trout. I heard John growl something to himself as a beaver splashed almost under the tip of his rod. As he whipped out some line in false casts, the beaver came up again about thirty feet away, and he laid the leader gently right across the top of its head with the fly a bit beyond. When he flicked the rod tip up to pick up his fly it caught in the beaver's ear, and instantly there was a great explosion of water as the beaver took off for its house. The reel screamed and light rod bent almost double as the beaver shot away. To try to hold a beaver on such tackle is about as effective as trying to hold a train by hooking onto the caboose. Then the leader broke and the line went slack.

That quieted things down considerably, and John was chuckling as he tied on a new leader and fly. No doubt the beavers were in their house—one of them wondering what kind of bug was fastened to its ear.

During the course of our fishing, John kept eying the beaver house speculatively, for some big trout were rising close to it. Finally he disappeared from view and when I next saw him he was wading out toward the house through the knee-deep water behind it. But when he got close to it, he found his way barred by a deep channel that the beavers had dug like a moat all around it—a sort of borrow pit for the earth they used in the construction of the house. Hanging his rod in the top of a drowned willow, John retraced his steps to shore, where he searched around through the brush till he found a dead log he could carry. Taking this back out to the beaver house he contrived to set it across the channel as a sort of crude bridge. It was an extremely unstable contrivance, being sloped and having one end floating, but his weight tended to stabilize it some as he took his rod and edged out toward the house again.

He was performing an astonishing feat of balance and was about halfway across when the log suddenly broke in the middle and he went into the ditch clear up to his chin. At that moment an extremely irritated beaver came up right in front of his face with its smallish black eyes bulging with anger and its teeth chirring in a most ominous sound. It must have sounded particularly bad to John, for his eyes bulged some too, but he prudently kept very still. For a few long moments he and the beaver eyed each other at a range of inches and then the beaver slammed its tail down and dove. John came out of the deep water about as fast as the beaver went under, expecting at any moment to feel big razor-sharp teeth fasten on some part of him. But the beaver was gone. While this episode has nothing to do with trapping, it does reveal something of beaver, one of the most coveted animals pursued by trappers. They are very mild and inoffensive animals, but when the young are threatened, or there is some untoward intrusion of their home pond, they will on occasion show fight.


INDIAN WAR PARTIES are no longer a menace. While trapping is an occupation not without risk, it is a healthy kind of life. Rarely does a man suffer from colds or any kind of infectious disease while running trapline, for there is no way for him to contact such infection unless some visitor carries it to him or he brings it in from civilization himself. Wilderness country is wonderfully clean, and cold germs and various viruses endemic to man do not exist there.

It is the unexpected accident or the moment of carelessness that is more apt to lay him low. Quite often danger comes so quietly and unobtrusively that a man scarcely recognizes its presence till it is standing over him.

I remember a day shortly before Christmas when I was about eighteen years old. I was running ten to twenty miles of trapline every day in snow a foot or more deep with temperatures running as low as twenty below zero. Having been on the trail for nearly six weeks, I was about as hard and tough as an eighteen-year-old can get—the kind of condition where the passing miles are scarcely noticed, and getting really tired was practically unknown. When I got hungry, I ate, and then was ready to hit the trail again. There is nothing quite so satisfying to a man as to be in that kind of physical condition, but knowing he is tough can also get him into trouble.

In those days every night of the week before Christmas saw a concert and dance in one or more of the small schoolhouses scattered across the country. These were more than just Christmas celebrations; they were a social institution. Families from each district with children going to school went to watch their offspring perform on stage and enjoy a lively social evening with their neighbors. The teenagers of the time thought nothing of saddling up a horse or hitching up a team to a cutter and taking in a different concert every night even though it meant going ten to twelve miles in winter weather. Most of them slept all day and danced all night during that week.

My trapline was yielding a steady catch of fur and I was reluctant to spring my traps during the holiday season. So at night I would come in to the home ranch, which was my headquarters that winter, eat a big supper, get all slicked up and dressed in my best, saddle a good big horse, and head out for the festivities. If the concert was not far away, I would snatch an hour or two of sleep before hitting the trail in the morning, but otherwise I just did without. It was a fair example of burning the candle at both ends. Along with the dancing, there was no telling how many miles I was covering every twenty-four hours. Things went fine for two or three days and the next night was no different from the rest except that the concert was about twelve miles away, and I didn't get home till after daylight.

Putting my horse in the barn and throwing him some feed, I got into my working clothes, ate a huge breakfast, and headed out along twenty miles of line. It was a clear, frosty winter day with the temperature hovering in the vicinity of twenty below zero. The snow was powdery and the traveling good. The snowshoes seemed to trail by themselves without me having to think about putting one foot ahead of the other; it was automatic. There was that good, comfortably warm, loose-muscled feeling that goes with being about as tough as a hunting wolf.

Noon found me in a little valley in a grove of heavy cottonwoods along the foot of the high mountains. I made a little fire and boiled some tea, and sitting on my snowshoes with my back to a tree proceeded to eat my lunch, relishing every mouthful, for I was ravenous. About the time the lunch was all gone and the fire about burned out everything went blank and I knew nothing for quite a long time. The sun was nearly set when my dog woke me up by licking my face and whining. I was cold to the middle of my bones, but except for a frosted nostril, not frozen. Dressed lightly in woolen underwear, wool pants and stag shirt, two pair of socks, and moccasins, I was not exactly rigged for sleeping in the snow at twenty below. Never have I been so utterly cold. If my dog had not roused me, it might easily have been my last sleep, for freezing to death and the lassitude of sleep are not too different. It is a creeping, insidious, and easy way to die. I just barely stepped back from the threshold that day.

That night I stayed home and caught up on some sleep.

There was another grim day that same winter, with the snow blowing on the northeast wind and the temperature well below zero. A bit before dark and miles from shelter I came down into the bottom of Drywood valley, where a log spanned an open stretch of creek. It was a big cottonwood laying on a slight slant and very slippery when covered with snow. Using a green spruce bough for a broom, I swept the snow off it as I carefully sidestepped across it. The creek was running with freezing slush and had dammed itself so that the water was three feet deep under the log—no place to fall into.

My crossing was made without trouble, but when my dog came across, he slipped and fell into the stream. The freezing slush grabbed his woolly coat, weighing him down and imprisoning him as effectively as if he had been tied with a rope. Cussing him, I went back out on the log, got down crosswise on my belly, reached down and pulled him out. Holding him with one hand till he got his feet under him and caught his balance, I turned him loose. Then I swung a leg up to get back on top of the log myself. But water dripping off his coat had frozen in a layer of clear, hard ice on the green bark, and it threw me into the creek up to my armpits. The water was burning cold and I was soaked to the hide when I crawled out.

There is no better way to freeze than getting one's clothes wet in that kind of weather—especially if there is any wind. Clothing turns iron hard in a few minutes and ties a man up like a straightjacket, making it almost impossible to travel. Knowing instantly that I was in big trouble unless I moved fast, I ran into a thick-growing grove of big spruces, got under one and pulled some fine stuff for kindling, got the match safe out of my pocket, extracted a dry match, and lit it. In no time a good fire was snapping and blazing cheerfully, while I peeled down to the hide. Wringing my clothes as dry as possible, I hung them around my fire.

Working on my underwear first, I got it back on before my wood supply was gone. Then with my moccasins on my feet I rounded up enough wood to finish the job. It was long after dark before I was able to step out on the trail heading for home. My match safe likely got me out of a bad spot that day, although woolen clothing can be freeze-dried, as was discovered later—but that is another story.

Nature itself can set some cunning deadfalls, sprung on man and animals with impartiality and hair-trigger, deadly speed. Of these the prime example is probably the snow avalanche. It can cut loose when least expected, set off by the vibrations of a rifle shot perhaps, or even less, and then thousands of tons of snow come sweeping down at express-train speed, smashing timber and everything else before it. If the snow doesn't get you, the terrific wind caused by air displacement when a large bulk of snow drops suddenly into a narrow valley likely might, and if this misses you, it is possible to literally drown standing up from breathing in too much powdered snow churned up by the mass of the fall. Everything living, large and small, caught in the path of an avalanche is in the gravest danger. One does not forget a close brush with one—the memory somehow lingers and induces cold chills up the spine.

There was an afternoon one winter when I was crossing a steep-pitched slope along one side of a big draw. I was almost to the bottom, coming down off the hill diagonally, when there was a sound like ripping canvas. In the fraction of a second it took to turn my head, the whole slope above me was a great mass of heaving snow chunks, some feet deep and yards across—a dread slab slide coming like a runaway train. Fortunately the soft snow was almost all blown away and my snowshoes were hanging on my pack, so that I could really run. It was only a step or two to the bottom and then bare slope rising beyond. Down and across the dip I shot with wings of sheer fright lifting my heels, and then up the other side. I just missed being buried alive. As the mass came to rest, a great snow chunk weighing at least a half ton just brushed the back of my leg.

WHEN A MAN CONTRIVES to make his living in the wilds he sometimes stumbles. And one of the most embarrassing things that can happen to a trapper is catching himself in one of his own traps, a misadventure that can be extremely uncomfortable and even dangerous on occasion.

One time Frenchy Rivière spent most of an afternoon setting a trap for an unusually cunning cattle-killing grizzly bear. With all the guile and detailed care of a master trapper setting a trap for a timber wolf, he buried and camouflaged the great fifty-pound steel trap, working on it till the ground cover gave nothing away to tell where it lay hidden waiting for the bear. Somehow Frenchy mislaid his ax, and while he was looking for it, he forgot where the trap was set and stepped squarely into it. Fortunately the trap did not break his leg, and ordinarily, with his setting clamp within reach, it would only have been a short time till he was free. But he had carefully hung the clamp in a tree well out of reach, so he was caught with one foot in his own trap waiting for someone to come turn him loose. The pain of arrested circulation and the pressure of the steel jaws and teeth were torture until the imprisoned limb grew numb. Fortunately, before any great damage was done, some of his family went looking for him and released him, more or less uninjured except for his pride.

Early one chill April morning years ago I was setting a trap for a beaver out in a dam along a deep submerged channel of the old creek bed. New ice like window glass covered the chain of ponds. The water was so cold it seemed to burn one's hide, and I could feel it chilling my legs through my waders.

The trap was a big, heavy forty-four and a half Newhouse wolf trap with a long, flat link chain constructed with a fork, snap, and ring with which a loop could be made to tie onto a stake or drag. Its loop was now hung around a dry stake driven deep into the muddy bottom on the edge of deep water. This set was designed to drown anything caught in it by the weight of the trap.

After setting the trap with a clamp, I placed it in position on the bottom and then proceeded to plant a couple of bait sticks of green poplar to attract the beaver. While doing this

I noticed that the position of the trap was not quite right, but before I got around to moving it, the clear water had become silted and cloudy. This did not worry me, for I was sure I knew exactly where it was, but when I reached down in the water to shift it things went wrong. There was a smashing jar, and I found myself almost to the top of my waders in freezing water with my hand hung up in my own trap.

The setting clamp was on shore behind me out of reach. The trap chain was held down by an inverted stub limb on the stake so the loop could not be slid over the top of it. There was only one thing to do: bracing myself, I took a short hold of the chain and began to work out the stake. About all I succeeded in doing was pull my boots deeper into the mud, until they were about to fill with water. Shifting a bit, I tugged at the stake from a different direction, gradually loosening it. But then I saw the top of the stake begin to settle and the hair lifted a bit on the back of my neck. The whole bank rimming the old stream channel was caving away into deep water and about to take the trap and me with it. Helplessly I stood and watched, my mind racing for some plan of escape, but then the stake came to rest again with its top about an inch under water. Gingerly I pulled on the chain again. Nothing happened, so I pulled some more. Finally after what seemed a lifetime, the stake came out and turned me loose.

Staggering out of the pond, half froze, with my waders full of water and wet to the waist, I headed for dry ground and my clamp. It was mighty awkward and desperate business trying to work it left-handed to free my trapped hand, but after some fumbling I managed to take the pressure of the heavy springs off the jaws. When the circulation started to come back into my hand the pain was sheer torture, making me grind my teeth. But after a few minutes it began to ease up a bit. Fortunately no bones were broken, but two of the teeth had almost met through the palm, although the skin was not broken. I knew then exactly what a trapped animal feels like, and the knowledge was illuminating.

LIKE MOST TRAPPERS I found the business of collecting wild furs adventurous, often hazardous, profitable, and about as free a way of life as is possible for man. A trapper does not fight the weather and the wilds, but learns to bend the elements and natural features of his surroundings to his use. He is at once an exploiter and a preserver—for he must exploit to live, but he must also conserve to be assured of a continuation of that living. Those of us who have abandoned the game look back with some nostalgia to clear rivers, clean air, and a definite memory of the kind of freedom that is not an embroidery of platitudes, but fact.

It is something to consider when a man contemplates the little red gods dancing in the flames of evening fires, and when he does, he is aware of some undesirable results of what we call progress. The principle of the professional trapper has always been to make use of a renewable natural resource without damaging the source, either by decimation of the wildlife from which he lived or by permanent damage to the ecology of his location. It is a principle that could well be put to further use.

8

# Tracks of the Hunters

MY EARLY AMBITIONS TO BECOME A MASTER TRAPPER AND hunter came to a nose-flattening halt when I first encountered the wily coyote. This fleet-footed, quick-witted gray ghost of the hills and mountains not only made a fool of me, but also reduced me to a size akin to proportions of a mouse in the relative atmosphere of competition between hunters. It was humiliating. It was shattering to my ego, which had elevated me in my own eyes as something superior among the life of the wilds. It was also illuminating, for such overwhelming defeat promotes thought, and through this something important was learned opening up some new and fascinating views.

In his own particular way the coyote is a writer and historian, as is most everything alive, including you and me. True, as with all four-legged ones, the coyote's writing is done in paw prints in snow or earth, and these last just as long as the elements of sunshine, wind, and rain choose to leave them unerased. These are the history of his doings for an hour or a day, for as long as they are left in readable form. And if a man is really interested in the way a coyote thinks and does things, he will cut a fresh track in snow early one morning and backtrack the little gray yodeler for as long as his curiosity and his legs hold out. For this is the way to read coyote "literature"—the story of his doings.

You will note I have said backtrack, for it is next to worthless to trail such an animal. Then results are only satisfactory up to the point when the coyote finds out there is a man coming behind him; after that the story dissolves in a long repetitious paragraph of feet being picked up and laid down very swiftly, a story of escape and distance which will go on for just as long as a man persists. Backtrailing will reveal an accurate and graphic account of how a coyote lives, what he avoids, and how he locates and stalks his prey, among many other things. In mating season it can also reveal a story of romance between him and his mate—the only thing missing being sound effects, which can be heard another time under the light of a full moon.

Whether it be tracks in new snow, fossils embedded in limestone, or the written pages of comparatively recent history books, backtrailing is a fascinating way to take advantage of hindsight.

If you have trailed bighorn sheep, as I have done countless times up over steep shelves and ledges of mountains in the Rockies, you may have seen their fresh hoofprints in dry dust drifted by the winds into the sheltering lee of a rock buttress—tracks so new their sharp-cut edges can be seen collapsing grain by grain of fine sand under the first soft brush of nature's erasing broom of air currents. If you choose to carefully whisk away the dust holding these tracks, you may well uncover more impressed in the underlying rock, the solidified marks of ripples matrixed on ancient beaches where wavelets lapped, and fossil remains of things living in an ocean millions of years ago—the vegetable and cold-blooded forebears of an awakening world, the very beginning for warm-blooded beings like men and mountain sheep.

Here in the Alberta Rockies at Waterton, not more than three or four miles from my door, can be found the fossil remains of an algae that once grew and flourished on the rocky coast of a great ocean, where huge breakers rolled and shattered themselves on what marked an Alberta coastline eleven hundred million years ago. This fossil algae, a cabbage-like growth called Collenia, was among the very first of all "living" things, an inanimate vegetable in an otherwise empty immensity, and now most exciting to a curious back-trailer. The only living counterpart on the face of the world today is a very similar algal growth found in intertidal pools between high and low tide marks along the coast of Australia.

By the same lure of curiosity a man's footsteps may lead him to a little town in France called Montignac in the valley of the Vézère River, where the cave of Lascaux is, the richest known prehistoric cavern adorned with man's earliest known form of picture art. Even more than this, it is the very beginning and the link between art and the vast world of communication—one of the very first records of transmission between human minds even though separated by many thousands of years.

The walls of the cave's galleries are a display of wonderful paintings some twenty thousand years old—a poetry of vivid expression telling us many things of the life of our predecessor, Lascaux man. We know from these paintings that he revered the animals he hunted, and this reverence may well be the root of all religion. Almost exclusively these pictures deal with animals: stags, bison, wild oxen, horses, ibex, and others parade across natural background landscapes impressed on the rough rock walls by sheer accident of water marks and geology. All are in warm color, alive and vital. One painting now named "The Man in the Well," due to its location, is particularly significant to the modern-day hunter and scribe.

This painting shows a bull bison with its intestines streaming from a hole under its flank. It is standing with neck bowed and horn cocked to gore and thrust, obviously taut with agony and anger. Ahead of it lies a dead man, frail and stiff by comparison, but obviously not responsible for the wound. For his throwing spear, lying to one side, is far too light and weak to inflict such damage apart from the fact that the wound's location does not go with such a weapon. Off to one side a rhinoceros is leaving the scene.

Without benefit of alphabet, the artist is telling us that a rhinoceros tangled with a bison, wounding him severely, whereupon the bull attacked and killed the man. Apart from the drama and color, this picture represents the first known hunting story ever set forth by the hand and craft of man. It marks the beginning of an epoch in human development wherein man worked to leave a record of his feelings and identity behind him, maybe aware of the fragility of his life and somehow appalled by the awareness, perhaps unconsciously trying to bridge the gap between him and the unknown shadows lying ahead by leaving his tracks in colors painted on stone.

The upper reaches of the Saskatchewan River were little more than trickles running down from receding ice masses of the last great ice age at the time this story had been adorning the walls of the cave of Lascaux for about five thousand years. As little as a hundred and fifty years ago, Stone Age man hunted here with strikingly similar flint arrow and spear points for bison, wapiti, and other animals. Man still hunts and camps across the width of mountains and savannas drained by this mighty river system, but now this activity is a foil for his fearfully artificial existence, a kind of struggle to recapture even for a little while the simplicity and reality of going out and killing his own food. As such, hunting is a very important part of man's life, a necessary portion of recreational activity enjoyed by fewer and fewer people due to waste, ignorance, and bad management. It is a kind of back-trailing relative to shrinking opportunities and the dismally unthinking destruction of the natural environment. It is an effort to remain, in spite of himself, at least in part a child of nature; as such, this search reveals the importance and reason for better management of our environment, without which the human species will ultimately perish.

Nobody knew better than the artists of the cave of Lascaux that if the hunt failed, they went hungry; by the same token, those of us lucky enough to be born and raised on the frontier, hunters from the time we could walk, knew the fierce joy of a successful stalk.

At the time my brother and I were old enough to press our bare feet into the soft warm silt of the river bars up along the headwaters of the South Saskatchewan, it was still frontier land, though tamer than what our grandfather found at the end of his journey west. But it was still vast, wild, and infinitely mysterious in our eyes. Hunting ran deep in our blood, a kind of unquenchable flame, a depth of compulsion to kill meat and gather skins fostered by our surroundings. Wisely our parents forbade us the use of guns till we were old enough and sufficiently strong to handle them safely. But this did not smother our desire for weapons. Nor did we develop any traumatic complexes or frustrations in the face of this prohibition. We simply accepted it and reverted to the primitive. If we could not hunt one way, we could hunt in another.

Our first weapons were catapults, the time-honored forked stick with two rubber bands and a pouch of leather attached, which many artists draw sticking out of the hip pockets of country urchins. Our first ones proved to be so crudely made that hunting with them was out of the question. But we improved them by buying special rubber bands made for the purpose, and carrying out considerable experiment with various kinds of ammuntion.

Small stones were only satisfactory for close-range practice shooting since there was considerable variation in size and weight. We broke up old cast iron into small chunks, which was somewhat better but still not sufficiently uniform for top hunting accuracy. It was not till we found an oldtime round-ball pistol-bullet mold of about .44 caliber that we really achieved the ultimate.

Being younger, John was of course smaller than I, but he used the same weight of draw and developed a sort of point and jerk technique that was deadly. We shot our first game with catapults—deadly weapons when loaded with lead-ball ammunition and fully capable of killing small game such as grouse, ducks, squirrels, and gophers. A ruffed grouse perched in a tree fifteen or twenty feet away was in grave danger of losing his life. A squirrel sufficiently optimistic to linger long at such range was likely to land in the bag. When Mother wanted a fryer out of the flock of chickens that ranged free in our ranch yard, she just pointed the desired rooster out to John and he would neatly behead it. When we were out exploring the mountains we often ate game killed with catapults.

While grouse would often stand for two or three near misses, ducks leaped into flight the instant they spotted us, which made them very difficult to kill. However, my first greenhead mallard was killed with a catapult, and the memory is still sharp, even today.

The drake was sunning himself on a half-submerged water-soaked log close to shore with two or three more of his kind, feet to feet with their reflections in the water, and somehow looking as big as geese to the small boy peering through a willow thicket thirty yards away. On hands and knees he crouched in the warm September sun facing the immensity of mountains across the lake, surrounded by dapples of light where the sun filtered down through golden aspen leaves overhead. Out on the lake the V of a swimming muskrat cut the still, mirrored surface as it swam past a small flock of bluebills, but the boy paid only passing notice, his eyes riveted on the mallard drake. His heart pounded against his ribs with excitement and impatience, but he had learned the folly of hurry in this game and he did not move a muscle for minutes on end while studying the ground between him and the ducks.

The slough grass was about a foot high and it was the only cover. With some luck it would do, but it meant belly-crawling over wet mud like a snake. The boy contemplated what his mother would say when he came home with muddy

clothes, the thought prompting him to slip out of his shirt
and pants, which he did so often for swimming that his skin
was as brown as an Indian's. With his loaded catapult in his
left hand and a spare ball tucked under his tongue, he
slithered out into the grass toward the ducks.

He had tried stalking ducks many times without success
but this did not dampen his desire to kill one; it only honed
his ambition and the way of his stalking. Sharp edges of the
coarse grass scraped at his belly, elbows, and knees, but he
ignored the discomfort, and when he slowly raised an eye-
brow to look, the sight of the big drake made his heart jump.

Closer and closer he crept over wet mud fresh-thawed
from the frost of the previous night, until icy water oozed
up around him making his muscles ache. When he dared
sneak another look the ducks were a mere twenty feet away.
A Susy duck, her rich-colored brown feathers freshly
preened, lifted her head with a sharp eye cocked in his direction. It was now or never. In one smooth motion the boy
came up on one knee with the catapult drawn clear back to
his ear. There was no time for a sitting shot, for the ducks
jumped with a wild clamor of quacks the instant they saw
him. The big greenhead drake seemed to hang still for an
instant with feathers gleaming and wings outstretched
against the blue sky. The boy's eyes were pinned on him as he
sent the lead ball on its way. Incredibly, the drake's neck
folded and he came plummeting back into the water, where
he lay on his back with bright orange feet pedaling futilely in
the air, as the boy charged out in splashing bounds to get him.

It was then he knew a feeling of savage ecstasy, primitive as that of a Stone Age hunter, the reward for his patience and endurance. Quickly he went back into the willows
and dressed and then proudly carried the duck to his mother.
So wild with his success that his tongue could barely artic-

ulate the words fast enough to tell about it, he blurted out the story of his adventure. She listened with proper awe to this tale of hunting prowess while admiring the drake. From the condition of his clothes and a glimpse of a splash of mud on his chest through the open front of his shirt, she guessed this half-savage son of hers had made his stalk as naked as the day he was born. Somewhat mystified, shaking her head in resignation but smiling at this business of boys growing up, she took the duck to prepare it for cooking.

Ducks did not come often to a catapult hunter, largely because of inaccurate pointing, not because the weapon isn't deadly enough. I found out soundly just how lethal a lead slug fired from a catapult could be and thereby learn the folly of mixing a hot temper with one's weapons.

We had a big black milk cow that went by the somewhat misleading name of Little Brains. If contriving trouble and various kinds of hellery can be attributed to intelligence, she was a genius. Only the fact that she gave a generous quantity of rich milk justified the trouble we encountered in getting it. From the moment she saw someone coming into the pasture to drive her to the corral to be milked, she would proceed to go through a whole series of maneuvers, from hiding in the brush to trying to bolt at the corral gate to finally trying to put both hind feet in the milk bucket even when hobbled. I hated that cow. I hated everything about her—even the shadow she cast and the tracks she made in the earth.

One evening when I went to gather the cows to be milked, I found the two of them in the thickest brush in the pasture. In the first two hundred yards toward home, the self-effacing Little Brains contrived to make a sneak off into the willows to hide, complete with her bell. One minute it was ringing somewhere ahead and the next instant it was gone. Using trailing techniques developed through long practice, I finally located her standing in a thicket, her bell silent. I tiptoed to within ten yards of her without being seen. She stood facing me with her face exposed and she made me mad all over just looking at her. Unslinging the catapult from where it hung around my neck, I slipped a lead ball into the pouch, drew the rubbers clear back to my ear, and shot her square between the eyes.

The result was electrifying to both me and Little Brains, for she dropped in her tracks exactly as though shot through the head with a rifle. There she lay, a great inert mass of the deadest-looking cow imaginable. I stood in horrified astonishment looking down at her, trying to think up some plausible story to tell my father and rejecting the idea just as fast. For it was perfectly obvious it would be most difficult to get him to believe even the truth. I just stood there caught in the snare of my own dilemma wondering what to do.

For once in her life Little Brains was cooperative and somewhat unconsciously generous. One of her ears came up to half mast from where it had been hanging like a broken hinge of doom. Then she gave a gargantuan gasp for air and began coming back to life. Looking decidedly drunk and anything but troublesome, she finally lurched to her feet to weave off toward the corral. She apparently associated this traumatic experience with me, for she proceeded to turn over a new leaf, never again trying to hide when she saw me coming.


WE TRIED BOWS AND ARROWS for hunting without much success, for we had no patterns from which to fashion such equipment. It was a lot of work, painstaking and very slow, to make a true, well-balanced arrow. An arrow is a very easy thing to lose in the heavy aspen and willow thickets where we did much of our hunting. We also had trouble finding suitable wood for bows, saskatoon being the best of the native varieties; but this normally grows stubby and crooked, so finding a straight stave of sufficient size and length was far from easy. Years later when I saw a saskatoon wood bow backed with rawhide on display in a museum, I realized the Indians also had their problems making good bows.

With all the impatience of youth, for whom time seems to stand still, we awaited the day when we would be considered big and strong enough to carry the responsibilty of firearms. When this golden moment finally arrived, it was a milestone in our lives, a kind of investiture from which we enjoyed a measure of bursting excitement and the much more sober realization of responsibility. Now we were a part of the loosely knit brotherhood of trappers, hunters, ranchers, and cowboys who carried arms as nonchalantly as they wore their hats.

To some, a rifle was just a tool to be used when necessary. To others, it was a sort of status symbol, and their guns reflected their pride of ownership. Among these were a small group of dedicated hunters and target shooters. These were the specialists, who not only owned more than one or two guns, but also knew how to use them. Most of them were deadly shots, who carried their interest to the point of loading their own ammunition and experimenting with various tools, bullets, powder, sights, and rifles. In contrast, there were men who thought so little of their rifles that neglect sometimes made their weapons virtually inoperable.

I remember being asked by an oldtimer to come to his homestead and shoot a crippled colt. I wanted very much to refuse, for I abhor having to shoot a horse, but those were the days before you phoned a vet to come with his needle.

A well-aimed rifle bullet was very sudden and merciful. Thinking the old man likely hated shooting a colt even worse than I prompted me to oblige. The colt had been born with incredibly crooked legs, grotesque in their deformity, and it never knew what hit it.

The old man was most appreciative and invited me into his cabin for a cup of tea. While I was visiting with him in his kitchen in a typical bachelor's abode, I noticed a rifle hanging on a couple of pegs over the door leading into his bedroom. Always interested in guns, I asked him about it.

He told me had taken it on a loan to an American cowboy who had come to Canada to compete in some of our rodeos. Somewhere along the circuit the cowboy had won this rifle along with some money, but ran low in luck at a poker game and lost most of his cash. He left the rifle as collateral on a small loan and then disappeared, never to return. The old man went over and took the rifle down, rubbed some of the accumulated grime of years off it, and passed it to me.

It was a Winchester of heavy caliber and looked as though it had hung there over the door untouched for years. It was coated with dirt and grease, but a bit of closer inspection showed it was essentially in good shape.

My curiosity aroused, I asked for some rags and kerosene, whereupon I began to clean up that rifle with the aid of a makeshift cleaning rod. I knew then why I had been asked to shoot the colt, for this rifle couldn't have been fired without blowing it up. It was plugged up with old grease inside—likely the original grease used in packing it for shipment from the factory. I also recognized with no little excitement what was coming out from under the camouflage.

It was a fine presentation-grade Winchester Model 1886 in .45-90 caliber, stocked in a half pistol-grip from a piece of beautifully figured Circassian walnut decorated with fine checkering and carved scrollwork on the fore end and butt. The sides and floor of the receiver were all magnificently engraved. The rifle was in mint condition with hardly a scratch on it—and by the look of it had never had a shot fired through the barrel. I got gloriously filthy cleaning it up till it glinted and shone. When I hung it back on its pegs over the door, it made everything else in the cabin look sadly beat up by comparison. Its owner sat, grizzled and unkempt, puffing on a huge, stinking bent-stem pipe, looking speculatively at his rifle and then at me from under the brim of his hat with eyes that had an amused glint in them.

Chuckling dryly, he remarked, "Never saw anybody get so damn dirty cleaning up somebody else's gun. Tell you what I'll do, seeing as it wouldn't take much to see you like that rifle. Give me fifteen dollars and it's yours. That is some less that I gave for it, but I never was much good with any kind of gun and it will only get dirty again if I keep it."

Being just a kid, I did not have that much money with me and times were tough. When I finally came back with enough cash to buy the rifle it was gone, for somebody had beat me to it with a price the old man could not turn down. To this day, nearly forty years later, I still look back with regret at not getting that magnificent rifle, for it was a jewel of the riflemaker's art created with the painstaking love of fine craftsmen for their work. That kind of skill and dedication is mighty scarce these days when planned obsolescence is an accepted thing among most manufacturers and the deep integrity of the oldtime craftsman is an attribute long gone and sorely missed. My mistake had been to clean up that rifle so that others could recognize its value.

. . .

OWNING A RIFLE and hunting with it was a satisfying thing, but for sheer excitement and spice of risk nothing could compare to following hounds on horseback in pursuit of coyotes. It was wild sport over plains country broken by draws and coulees along rivers where coyote hunting was best, and nobody ever enjoyed it for long without knowing what it was like to take a hard fall. With snow on the ground in winter, the ever-present badger holes were a particular hazard, for under the white camouflage the holes were hidden so even the best mounts could not spot them in time. Broken bones were a hound man's hazard, and the coyote was by no means always the loser. Even the dogs did not always get off without casualties. Long before I was old enough to hunt with a rifle, I rode with hounds.

Running coyotes took big, fast dogs—fast enough to catch a very fleet animal and then big and tough enough to make a kill. Coyotes rarely weigh more than about twenty-eight pounds unless glutted with fresh meat, when they will often weigh ten pounds more, but every ounce of it is canny stamina and fight. Popular belief tags the coyote as a coward, but nothing could be farther from the truth. He is just too intelligent to fight unless cornered, but when the chips are down, the little gray yodeler can fight like fury and many dogs wore scars to prove it.

When hounds first appeared on the prairie, almost any kind of dog that could make a decent run could catch a coyote, for they had never been hunted like this and were at first confused and at a loss to cope with dogs. But soon all the slow coyotes were gone and only the fastest and canniest of the breed were left to propagate the species. Naturally selective breeding had its effect, which was good for coyotes in general and caused dog owners to go to considerable lengths to improve their hound packs.

On flat ground where a coyote did not have more than two or three hundred yards' start, the balance was well tipped in favor of a good hound pack making a catch. But if the quarry was jumped close enough to rough ground or broken, hilly country, the odds were in favor of the coyote. A coyote can out-dodge hounds on steep ground, for its center of gravity is lower and it is short-coupled, allowing for faster turns. On such footing hounds nearly always end up laming themselves, cutting and tearing their paws on gravel, prickly-pear cactus, and frozen ground.

Coyotes with experience never miss a chance to take advantage of any feature of environment in their favor and often make lightning-fast decisions while running at top speed—decisions timed to the split second, with never a step lost in execution of a maneuver.

A pack belonging to one of my grandfather's neighbors ran a coyote out onto the ice of the St. Mary's River one day, where the wise little speedster contrived to put a strip of open rapids between himself and the pursuing dogs. Such hounds are strictly sight runners, and so keen are they that when close to their prey they are almost oblivious to all else. Cutting across the arc of the coyote's circle to get at him, the whole pack piled into the fast water, where half the dogs were swept away under the ice to their deaths. There were times when the price of coyote fur came high.

I remember seeing a pair of good hounds belonging to my uncle Ernest run a coyote into a four-wire fence, where it turned to go threading back and forth between the posts under the bottom wire. Obviously the coyote had used this technique to escape before, for the whole maneuver spoke of practice. Many hounds would have cut themselves to ribbons in a spot like this—especially a pack of several dogs where competition runs high—but this pair of dogs was accustomed to running as a team. The big lead dog, a very fleet animal, jumped the fence in one long flowing graceful bound, turned and flanked the coyote on his side, while the bitch came up on the opposite side. They almost caught the coyote, but about two steps ahead of a violent end it dropped into a dry wash leading steeply down into a brushy draw and disappeared.

Hounds live to run; the hunting horses involved in such chasing also grow to love it, and like the men in the saddles they sometimes let their enthusiasm overcome better judgment. Excitement bubbled in the blood as hoofs drummed and eyes were all afire with action.

One winter when I was about ten years old, I was visiting during Christmas holidays at our grandparents' ranch and somehow managed to talk my uncle into taking me on a coyote hunt with the hounds. He mounted me bareback on a fast little brown mare that had shaggy hair all over her as thick as bear fur and a long woolly mane and tail. There were two or three inches of new snow on the ground, making everything sparkling bright under the morning sun and it was just cold enough to make a man and a horse want to go for a run.

The horses were prancing along behind the hounds jingling their bits and tossing their heads with impatience for action. The hounds were a pair of handsome crossbreeds, the dog being the bigger and faster, the bitch being the killer. He was named Weasel and she was Brindle.

Weasel and Brindle had made many a successful run with a system that worked very well for them. Because he was slightly faster, Weasel was the catch dog. He would run up on a coyote and upset it end over end with a snap of his jaws and quick toss of his head. This gave Brindle a chance to close up the lead and before the coyote could get going again she would fasten on its throat to make a quick kill. Weasel was generally phlegmatic once the catch was made, and was perfectly content to stand aside watching the enthusiastic Brindle administer her efficient mayhem. Only if the coyote was big and fighter enough to make trouble for her would he fasten his teeth on a hind leg and throw his weight sufficiently into the mixup to stretch out the coyote.

This particular morning, we had not ridden more than a mile from the buildings up onto a bench along the valley, when we suddenly jumped a coyote out of a weed patch a couple of hundred yards ahead of us. It broke and ran across a flat pie-shaped piece of high prairie pointing into the junction of Pothole Creek and the river. Both dogs sighted it instantly, and went flying off in pursuit with that long beautifully flowing stride so typical of running hounds. In two jumps my horse was going full speed right behind them, as though running free, which indeed she was, for I weighed scarcely eighty pounds. I rode crouched low over her withers, one hand clutching a bunch of her mane and the other holding the loose reins. In spite of a protesting yell from my uncle, we shot away from him and my hand did not move to check the horse with the bit, for to do so would have been an invitation to be thrown. I had learned to ride bareback in mountain country, knew excitement when I felt it under the seat of my pants, and had no intention of pulling myself off the mare's back trying to slow her down.

Peering through a wind-torn storm of flying horse hair with the wind tearing at my ears, I rode like I had never ridden before, listening to the drum roll of my horse's hooves as she tried to catch the hounds. My cap went sailing away behind. The hounds were slowly pulling ahead as they cut into the gap between them and the fleeing coyote. It was wild and wonderful—fast action out there on the snow-

239

blanketed prairie under the big blue bowl of winter sky—
the kind of thing for which boys are made, and spine-tin-
gling to contemplate even in retrospect.

For half a mile the going was as flat and smooth as a
billiard table with excellent footing for the animals. The
dogs were about a half dozen lengths behind the coyote and
twenty lengths in front of my horse's nose when they suddenly dropped from sight like magic. One second I was looking at them between my horse's ears and the next instant
they were gone. And then I was looking almost straight
down from the brink of a drop-off into a steep-pitched draw
running down a quarter mile or so toward the river flats
below.

My horse and I were almost airborne as she took the
slope. All that kept me from leaving her completely was my
grip on her mane, for the seat of my pants was floating on
inches of daylight as she slid and bounded. She went with
her head low to see where she was going and her haunches
tucked under her in long sliding bounds straight down. I
caught a glimpse of the hounds and coyote barreling almost
together in a flying cloud of snow, skidding and twisting
into a big washout away down the draw.

By some miracle of horseflesh and balance my horse and
I managed to stay right side up. Almost without a jar she
came to a stand with the hounds and coyote fighting right
under her nose. There was a flurry of gleaming teeth and
snapping jaws so fast it was hard to follow with the eye. Then
Brindle got her hold on the throat and Weasel hauled back
on a leg to help her, and they made the kill. Both dogs, the
horse, and I were panting for breath, drunk with excitement
and oblivious to all else but the action. It never occurred to
me to be scared, and it was not till my uncle showed up rein-

ing his horse down the steep slope that I realized I was out of favor. Giving me a sharp look and muttering something about a damn-fool kid trying to break his neck, he got off his horse to pick up the coyote and tie it behind his saddle. Then we headed for home.

Because I rarely visited the prairie ranch in winter, I never got to ride with the hounds again in Ernest's company —a fact always regretted; for he was fun to be with and this was about the most exciting kind of hunting ever experienced. However, I did get to do some jackrabbit hunting with a hound one summer, and once this hunting got sidetracked with salutary results.

Somebody sold Uncle Ernest a big, raw-boned female hound pup that looked like a cross between an Irish wolfhound and a Scottish staghound, with perhaps a dash of greyhound thrown in for good measure. She was tall with a brindle-colored rough coat, and at first very awkward; but as she grew she showed promise of great speed and stamina. We called her Whiskers, and as a small boy I loved to take her and a fox terrier out into a cornfield by the ranch buildings to hunt the jacks. A hound could never find rabbits in cover like that, but the busy terrier's nose was sharp and she gloried in sniffing them out. Once a jack was jumped from its hide it was fair game for the hound, and if she could close before the long-eared speedster reached the hills running up from the river flats, she had little trouble making a catch. It was a picture of fast action and cooperation between two dogs about as far apart in size and disposition as dogs can get.

While Whiskers showed promise and speed in her ability to catch the big hares, it was very apparent she had a built-in antipathy toward coyotes. Their howling at night launched her into spasms of trembling and the sight of one was enough to send her streaking for home with her tail between her legs. She had an accident one day that did not improve her attitude.

I had her out in the cornfield hunting jacks, and in due course the terrier jumped one out of the corn, which immediately set out for the hills at high speed. Maybe this one had been run before, for it did not linger for any second looks and it was helped by Whiskers' failure to see it until it was well on the way. But in spite of its lead, the hound made a terrific run and had the range cut to a few yards when the rabbit came to the foot of the slope. Up it went like a streak, and as luck would have it, almost ran over a half-grown coyote pup sunning itself not far from the den. The next moment the coyote pup found itself confronted by the hound. As it fled for the hole somewhat demoralized by the sudden danger, it overshot the front door with Whiskers' eager jaws snapping just short of its tail. The hound had things mixed up a bit and proceeded to pursue the little coyote instead of the rabbit. Around and around they sped, the steep pitch of the slope saving the coyote's life about twice on every circle, for the hound's excitement and eagerness were causing her to fall over her own feet.

At this point mother coyote showed up and promptly joined the procession. After a few moments she got close enough to Whiskers to snap her teeth on the handiest part of the hound—her long stringy tail. Whiskers was understandably amazed to find something biting her vulnerable rear, and when she looked back to find a full-grown coyote attacking her, she came apart at the seams. Howling and crying, she stampeded for the buildings. When she got there, she crawled under the kitchen steps and stayed there for most of two days too frightened even to eat.

Uncle Ernest became so disgusted with her that he gave

her away the following winter. Her new owner tried running her with his pack. This seemed to do something for her, and Whiskers developed into a top-notch catch and killer dog—a rather rare combination of talents in a coyote hound. But her illustrious career was cut short prematurely, for she was one of the dogs previously mentioned that got tolled into the open hole in the river and drowned.

By the time I was running trapline along the foot of the Rockies to the west, my interest in coyotes surpassed the excitement of pursuit, for then I was after skins, which meant money to be converted into the necessities of living. I pursued coyotes with snares, traps, and rifle, and it was through the use of these that most of my experience with coyotes was gained. By the same route I developed an eventual antipathy for any kind of trap and a lasting appreciation of fine firearms—an appreciation that has never waned.

SO THE LITTLE GRAY YODELER became my teacher—an unwitting one to be sure and most certainly an unwilling one had it been aware of its contribution, but nonetheless a most fascinating instructor with an enormous talent for sharpening one's powers of observation, patience, and physical stamina. Never did the student become so filled with avarice that he stooped to the use of poison, and never did his education become so complete that the coyote population was decimated in any portion of the range. We were associates sharing a big piece of rough country with an abiding respect for each other. We played a game as old as time, where a hunter pursued a hunter, each using every trick and wile to win. It was a wild, primitive game often played in magnificent surroundings, where snow-draped peaks shining under a winter sun looked down on these moving specks on distant slopes—we were part of the wilderness, wonderfully free and sometimes breathless with the sheer excitement of the playing.

What is life if it is not sometimes free of the shackles called civilization? For only under the vast blue vault of open sky can a man come to know himself and the true values of his association with other forms of life. It is not in artificial mores and rules that life is really lived, but in following the laws of nature. So spoke the coyote to me many times—sometimes vocally under a silvery moon when frost cracked and snapped in the snow-shrouded timber, sometimes in silent sign language; but the message came through strong and clear like a wild ballad accompanied by the sound of wind in pines or the profound silence of a winter's morn when the peaks stood tinted in rose and purple shadows in the light of the rising sun.

9

# *More Hunting Trails*

ONE CRACKLING COLD WINTER DAY I WAS OUT HUNTING coyotes up along Pine Ridge, a high watershed overlooking the Waterton River valley a few miles below the lakes. There were about six inches of new snow on the ground and it was still snowing a bit, with fine stuff drifting in on a northeast wind that felt as though it had come non-stop from the North Pole with all the trimmings that go with a draft from that end of the earth. A good way to keep from noticing the cold too much was to ramble on foot with a rifle. Tracking was good and along about mid-morning I struck a coyote track that was paint fresh.

About ten minutes later I came to a spot where the coyote had stopped to feed on the guts of a deer left by someone who had cleaned out a buck two or three days before. The tracks that left this spot were even fresher, and I knew I was getting close to the animal that made them. Sure enough, after going another half mile or so, the coyote jumped out of a brush-filled draw in front of me and streaked away out along the open slope of the ridge. Flopping down on my belly I picked him up in my sight and followed him with it, hoping he would stop for a look back to see if I was coming. About two hundred yards away he checked on a steep little hogback ridge and turned broadside. My rifle cracked and the coyote dropped in his tracks, shot clean through the heart—another nice fur to pull over the stretching board and another box of groceries for the family.

There was nothing particularly notable about this incident, and likely it would be lost in memory among scores of other hunts if it had not been for what followed that day.

Along about noon, I was swinging back toward home and was crossing the road at the west end of the ridge when a neighbor, Eddie Schmidt, came along with his truck. Eddie is a big man with a built-in sense of humor and one eye. He and I hunted a lot together in those days and it was amazing how quick he could spot game and how accurately he could shoot in spite of his handicap. He saw more with one eye than most men ever did with two and was a dead shot with his rifle. When he stopped that day his sense of humor was not very noticeable, for he told me a bunch of elk had broken into one of his stocks of winter feed and had done a pretty fair job of wrecking it. He wanted me to go with him to shoot a couple in hopes of scaring the rest of the bunch back up into the mountains of the Park. This seemed like a good idea, especially since neither of us had killed our winter's meat as yet.

I had already been over eight or ten miles that morning and had had no food since breakfast before daylight, but rather than take the time to go for something to eat, we just left the truck parked along the edge of the road and headed out across country for a big basin up on the head of Cottonwood Creek. Eddie had trailed the elk far enough to know they were headed in that direction, and we figured they would likely bed down for the day among some heavy willow thickets on a frozen swamp where most of the creek came from underground in a series of big artesian springs.

After about an hour's walk we came out on the crest of a low, open ridge between Pine and Cottonwood Creeks from where I could see out over the swamp with my binoculars. A quick look showed elk scattered everywhere. There must have been close to a hundred head of them. It was no wonder Eddie was concerned about his winter cattle feed, for when a bunch like this breaks into a haystack they wreck it in short order. In those days, before the advent of modern power hay balers, we stacked our hay loose and fenced the stacks. No kind of ordinary wire fence suitable to hold domestic livestock would turn an elk. Feed is a vital part of a livestock operation during Alberta winters and nobody could afford to play host to a bunch like this very long.

The weather was not improving, and by this time fine ice crystals were driving on the wind, making the mountains at the head of the basin look as though they went clear to heaven before topping out. Their outlines were huge and ghostly without much detail under the pale light of the sun through the frost haze. To get the wind right we circled away wide to come into the willow swamp with the breeze blowing square in our faces.

We went very slow and careful through the thick willow growth, stopping often to listen, and after about a half hour or so of this we began to hear the elk snapping brush here and there, giving those typical conversational squeals as they begin to browse. We could not avoid making some noise in the thick stuff, but this does not matter much when working up close to a bunch of elk if a man stops often and takes lots of time. For there are always noises among a herd and as long as a man contrives to sound like another elk moving through the brush, they pay not much more than passing attention.

Eddie was ahead of me on a faint trail and when we came to a place where it forked, we decided to split up. I had not gone more than twenty steps when I saw a cow elk staring fixedly my way over the top of a clump of willows. We had agreed that whoever saw an elk first was to shoot. That way there was a better chance of one of us making a sure kill before the elk took fright and stampeded. So I lifted my rifle to bring my sights to bear squarely between her eyes and squeezed the trigger.

The shot set off action all over the place. The cow I shot at disappeared as though she had fallen down a well, and brush began to break as the herd began to move out. Then Eddie's rifle crashed once and I knew another elk was down. Upon going to the place where my elk had been standing I was surprised and somewhat disgusted to find it laying stone dead squarely in the middle of a sheet of ice frozen over one of the springs that beaver had dammed up into a pool about a hundred feet wide. Ordinarily these springs did not freeze, but the damming had allowed a bigger surface to cool and now a thin layer of somewhat treacherous-looking ice covered the deep pond. I liked nothing about the look of the place, but then I realized that what would hold up a cow elk would likely hold me up too. To make doubly sure I began circling the edge of the pond to the place where the elk had left the shore so I could follow her tracks out onto the ice. My way took me over a string of hard snowdrifts that the wind had piled in the willows after some previous storms. Without thinking about it I was about halfway when I walked square into a natural trap.

When the beavers had come to this place they had not only raised the water level but also dug some deep channels back into the willows. Where the snow was piled over one of these on the rim of the pond, it looked no different on top, but underneath, the warmth of the water had undermined the drift and softened it. Suddenly, without any warning, like a trap door dropping from under my feet, the snow crust let go and I found myself standing up to my armpits in icy water looking up at the sky through a hole three feet over my head.

It was one of those times when a man finds himself in a spot where he has to make some fast moves and they had better be the right ones or he is apt to end up very dead. Yelling would not do much good, for the snow would muffle it and if I waited till Eddie came to investigate I would likely be too cold to care any more. Anyway, the biting cold water had taken my breath away to the point where yelling loud enough seemed utterly impossible. Moving might drop me into deeper water but I had to take the chance. Using the butt of my rifle I dug into the snow ahead of me, undermining it till it fell in on top of me, then squirmed ahead through the slush to dig some more. Finally I uncovered a ledge of thin ice and frozen mud strong enough to hold me up, and from there I was able to crawl out.

But I was soaked to the neck and miles from home with the wind blowing—its chill registering somewhere near twenty-five below zero or lower. Instantly my clothes began to freeze, and I knew I was in big trouble if something was not done quickly to get at least a part of the water out of them. There was little fuel handy for a good fire, since most of the willows were green. Kicking the snow off a big hummock of dry grass, I proceeded to strip, wringing out my clothes as I took them off. Fortunately I was wearing pure wool from head to foot or I never would have dared attempt it. My plan was to wring as much water out of my clothes as possible and try to get home before I froze.

My socks came first and were put back on my feet. Each item of clothing was left in a bunch so it would not freeze immediately and stacked on a dry snag as it was wrung out. By the time I was working on my underwear, dancing around naked and swearing under my breath at the burning wind, two things happened close together. First, the snag holding my clothes broke and dumped all my wet garments into the powder snow, which instantly stuck to them. Although I did not see him immediately, Eddie chose this moment to look out of the bush onto the little flat where I was jumping around.

Mad all over and plumb disgusted at things in general, I picked up my undershirt and swung it hard against a willow to knock the snow off it. Taking it out on that willow helped a little and the pounding seemed to knock the frozen moisture out of it. After some vigorous beating I put it back on, sure that it was a whole lot better than nothing for if it was not completely dry, it was also a long way from being wet and besides it broke the wind. Grabbing my long-handled drawers, I proceeded to swing them at the willow, and about the time I was set to pull them on I heard a funny noise and looked around to see Eddie staring at me with about as comical a look of sheer disbelief and amazement imaginable.

"What happened?" he asked. "You gone nuts or something?"

"Hell no!" I snapped. "I shot an elk and I'm celebrating!"

By the time I got my clothes back on and found a way to get out to my elk, I was beginning to warm up. It did not take very long to clean out the cow and prop the carcass open to cool, but it was after dark by the time I pushed open the door at home, and the temperature had dropped another ten degrees. Never did stew bubbling in a pot on the stove smell so good, and while I was getting into some dry clothes, Eddie regaled Kay with the story of our hunt. She was looking at him as though she thought he was second cousin to Baron Munchausen when I came back to the kitchen, and he was laughing all over as he accounted how funny I looked dancing the prairie-chicken dance out there on the frozen swamp stripped to the hide. About then, with a drink of good whisky and a hot plate of stew in front of me, I could see the funny side of it myself, for apart from a frost nipped heel I felt fine.

IN OVER FORTY YEARS of intensive hunting, exploring, and wilderness experience—over countless miles involving boats and whitewater rivers, broncs in mountain country, wounded bears and foot-slogging on snowshoes, skis and plain shank's mares; my closest shave to getting a one-way ticket to the cemetery came while camping one summer up on Boundary Creek in Waterton Lakes Park, hunting big-horns with a camera.

This region ranks high among the grandest mountain country of the world. It is a land of color, where the rock formations of the mountains run the spectrum from green to red to purple and a multitude of shades between. Our private campground there, hidden among the spruces and alpine larches by a clear, cold creek winding across a flowering meadow, is like a playground of the gods. It is surrounded by country teeming with big game of all kinds and includes some of the finest summer ram pastures in the Canadian Rockies. Here and there lakes are pocketed among the peaks, rough-set like sparkling jewels and some of them full of fighting trout. It is heaven for a summer wanderer of wilderness country—especially a camera hunter.

We burned wood in our folding camp stoves, and one brilliant August morning I took my horse and a four-foot Swede saw a quarter mile down the creek to where it dropped over a cliff on the edge of a piece of old burn. I was after a suitable dead tree for firewood and this was a good place to get a bone-hard fire-killed log of a suitable size to drag back to camp with my lariat.

Some of the standing dead timber was too big to handle, but I soon found a snag about twenty feet high and perhaps twelve inches through at the butt. It was standing right on the edge of the drop-off a bit to one side of the falls, where the creek made its first thirty-foot jump toward the valley floor far below. It was a tricky place to fell a log, but after looking it over from all angles I figured it could be made to drop uphill if properly undercut.

So I proceeded carefully, taking lots of time with the sawing, and when the snag began to tremble and crack its top was swinging exactly as planned. Clearing the saw blade, I stepped quickly to the side onto a deadfall lying in the waist-high brush—a safe spot in case my log jumped the stump. That step came near being my last. For the log on which I stood was well hidden in the thick snowbrush, and it was impossible to see that it was lying over another log—a natural fulcrum—with its broken-off tip projecting across the spot where my log was due to land. The snag came crashing down and the next instant I was hurled straight up and out in an arc toward the edge of the canyon below the falls. It was exactly like being flipped off a springboard and I was plenty high enough to have lots of time to think.

I have heard said that in such a spot a man's past life flashes across his mind, a sort of instant review of wherever his trail has led him. I am not inclined to accept this as fact—certainly not in my experience—for any time a tight spot and possible violent end have confronted me, my reaction is the exact opposite. It is like a bit of slow-motion picture action, wherein every detail is vivid and sharp, although any moves on my part are usually purely instinctive.

Now I was airborne and my only thought was that I would probably be able to hear my bones break when I came down. At the same time I somersaulted completely in the air, putting my back to the drop-off. There was a branchy second-growth fir tree perhaps twenty feet high growing up from the edge of the canyon and I lit astride it a few feet down from the top. My weight made it bend out over the edge in a great bow, where it solemnly hung for a moment before starting to straighten up. I didn't lose my balance for a moment, and never felt a jar. I just came slithering down over the branches feet first to land gently without so much as a scratch. Had it been rehearsed and practiced a thousand times, it could not have worked out better. It was the most fantastic sequence of unfortunate and equally fortunate circumstances anyone could possibly encounter. It was one of those times when a man stands in awe wondering how come he is still alive and unhurt—not quite able to believe it. My pipe had been in my teeth while working and now I found only the bit in my mouth, for I had bitten it square off.

So what comes unannounced can be dangerous, comical, interesting, or even fascinating, and no man who follows outdoor trails can overlook that possibility. For if he does, he may well relegate himself out of the realm of sport into the world of sheer mechanics, where he becomes the kind of "sportsman" who cruises country roads with the windows of his car screwed up, the heater turned on, the radio blaring, and his modern high-powered rifle beside him, while he watches for the opportunity to murder a deer or some other kind of game with a minimum of effort. He knows little about living in the open and much less about trailing a wounded animal. He, after all, carries a rifle of such awesome ballistics that wounded game is thought of as being other people's headaches. As one old mountain guide put it, this kind of man carries a gun big enough to shoot a track and kill the animal that made it. If he has the bad luck to get caught out overnight in a blizzard, he is likely to freeze to death. His manners are bad and his ethics worse. There is really nothing wrong with him except that he is a killer knowing nothing and caring less of the finer points of the game. He is a kind of powder-burning bingo player who stacks the deck against the wildlife out of sheer lack of recognition of his true relationship to other forms of life. He is licensed as a hunter, a form of legalized mishandling of the possibility that he may not know the muzzle of his weapon from the butt plate, and may be on his way to killing another

man—accidentally of course—for, after all, he is duly licensed to hunt wild game. One does not speak of this man
with complete derision but with some pity, for he does not
know what he does or what he is missing.

REGARDLESS OF WHAT a man chooses to use in pursuit of sport,
be it gun, fishing rod, or camera, or how well he uses it, the
day is enhanced by the unexpected. It was this way the time
Ed Burton's hat got mixed up in the bighorn hunt. Like
riding into a hornet's nest, it started a whole train of un-
forseeable events. Because of the hat the eagle showed up.
Its arrival impelled me to move down the ridge to see if Ed
was still in one piece. Then I climbed out on a point of rock
to have one more look into a basin that had appeared completely empty of bighorns for ten interminable hours. That
look was enough to change the whole outlook and atmosphere of the hunt.

Ed and I had planned this hunt months in advance. He
was a rancher, rodeo contestant, cougar-hunting specialist,
my brother-in-law, and a man who would sooner hunt sheep
than eat. He is gone now, but he left a legacy of many
memories.

Although both of us had lived in bighorn country all
our lives, neither of us had ever taken a really big ram. An
old game preserve of forty years' standing was being opened
for hunting that fall of 1954—an area located in the Castle
River drainage among the peaks of the southwest Alberta
Rockies. There were bighorns in this region and with a little
luck we hoped to collect two good trophies.

Two days previous to the opening of the season Ed arrived at my headquarters in Waterton Lakes Park. I was

outfitting and guiding for a living then, so this was to be a sort of postman's holiday for me, as well as being a sort of short opening round for a busy hunting season.

When we were packing up the horses with our grub and gear, Ed found to his disgust that he had forgotten his hunting cap; so when we went jingling off up the trail for high country, his head was adorned with a white ten-gallon Stetson. And, as somebody once said, thereby hangs a tale.

We camped that night just under the Continental Divide beneath a towering, almost vertical wall of rock miles long and reaching well up to two thousand feet in height in places. The tent was pitched on the edge of a meadow where ice falls and avalanches every winter and spring kept the timber from growing over a strip of ground like a giant's bowling green of lush green grass. A little behind the tent directly under the cliff, a small mountain tarn was set among the rocks where falling ice and snow had gouged away over the years to carve a natural bowl that held the water. It was a beautiful, wild, and uncompromising place in which our tent added a touch of cheer, shining with the light of candles inside as night settled over the mountains. Echoing off the cliffs came the music of the Swiss bells of our horses as we rolled into our sleeping bags. At intervals during the night we woke to the sound of rocks falling off the precipices above, rumbling and clattering in dizzy plunges until finally coming to rest on the talus fans below. It was a common enough sound to those accustomed to camping among high mountains and no menace, since the tent was well out from the foot of the slide rock.

Well before sunrise next morning we were in the saddle heading for the top of a ridge a mile or so by crow wing to the east. We tied our horses in a larch park just under timberline not far from its crest. This rim overlooked a huge steep-walled basin at the head of a southern tributary of Castle River, a wild piece of precipitous country that I had first seen as a small boy rambling among the peaks with John on our cayuses. Now it was an ideal spot to scout for sheep —we hoped to collect a couple of rams on the following morning, the opening of bighorn season.

Ed had taken some good-natured ribbing about his flamboyant headgear, and I couldn't resist remarking, "You're in tough shape for a sheep hunt! Wear your hat and the sheep will see you for miles. Take it off and they will sure see the sun bouncing off that bald head twice as far. You'd better crawl under a clump of brush before you peek over that ridge."

"Just wait and see," Ed grinned. "This war bonnet is lucky."

As we headed for the rim we split up. I went straight up the slope while Ed slanted away down along its spine for about a quarter mile before topping out. When I came out of the trees up onto the rim, I spotted him crawling on hands and knees into a clump of shintangle fir, where he disappeared completely. He was taking no chances in his choice of a lookout. I bellied into a comfortable niche among some rocks and cautiously lifted my head for a look.

A breathtaking view of the great basin was exposed. Right under my chin the cliffs dropped away in dizzy plunges to the top of the rock slide a couple of thousand feet below. For at least a mile both ways from my perch the rock wall bent in a giant arc, the lower shelves, slides, and timbered crannies being ideal range for wild sheep.

The binoculars revealed a network of fresh game trails crisscrossing the rock slides in every direction. My nine-power glasses picked up a dozen goats—a regular convention of the white, whiskery climbers of the crags. Some were bedded down, and some were feeding on the cliff faces in the usual impossible places frequented by these animals. Slowly combing the ledges, slides, and gullies, I worked the glasses over the basin, but no sheep of any kind showed up.

The sun came up to climb across the blue vault of a cloudless sky—hot, bluebird weather and not the best for hunting but warm and pleasant—much more so than the wind-driven snow and rain often encountered in high country in the fall. Constantly my glasses probed and roamed. There had to be sheep here somewhere—I could almost smell rams—but none showed up. My lookout commanded the whole basin except for some timbered shelves directly under me, but it was likely Ed had a view into these so I didn't move.

Noon came and passed. The afternoon wore on ripe with expectancy: there had to be sheep among these rocks. But none appeared. My eyeballs began to feel like two well-fried eggs, so I rolled over in the deep shade of a rock to give them a rest. I was dozing when I was suddenly jerked back to consciousness by the unmistakable thrumming roar of a stooping eagle. At the same moment I spotted the big bird dropping in a sizzling dive with wings almost closed straight toward the clump of brush where I had last seen Ed. Scant inches over the tangled wind-twisted scrub, the big golden eagle checked to shoot straight up a hundred feet, roll over on his back and strike back toward the brush again. This time he grazed the evergreen tangle and I saw the top of a small tree jerk and the flash of a white hat. The eagle must have spotted his mistake then, for he peeled off over the cliff to soar away.

My astonishment gave way to suppressed laughter as I crawled out of my hide to swing down through the timber toward Ed. He met me, still wearing the white hat and an expression of surprise.

"What do you suppose that damn-fool bird thought it was doing!" he exclaimed. "He came within inches of scalping me!"

"I'll bet that eagle was just as surprised as you were," I told him. "He figured he had a kid goat or a rabbit spotted and it must have been some shock to find out he was about to grab a tough old sheep hunter."

Hunkering down among the larches, we held a council of war. Ed had seen nothing but goats, but he too was convinced there had to be sheep not too far away.

"Offhand this basin looks like a bust," he opined, "but I'll bet there are rams around here someplace. Maybe we'll find them tomorrow."

"Before we pull out for camp I'm going to have another look," I told him. "If there are rams hidden somewhere in this hole, they should be getting up to feed soon."

Just above us a rocky buttress thrust itself out from the ridge top and I crawled out on this to give the basin another thorough glassing. There was still nothing in view but goats, but I noticed the still-concealed shelves under my former perch were better exposed, so I put the glasses to work on them.

At least a thousand yards diagonally down the rock face something caught my eye at the foot of a gnarled old spruce. It looked like a weather-beaten root, but something about it made me curious. Snaking out along the top of the buttress a few feet farther I put the glasses on it again. The root had not moved. It was the right shape and the proper slope for a root, but its color puzzled me; so I kept the binoculars pinned on it wishing for the spotting scope left at home.

Then there was a tiny flicker of movement up higher along the bole of the tree, and like magic the curve of a big ram's horn was revealed. The odd-colored root was the ram's hind leg as it lay sprawled and almost hidden behind the tree.

When I signaled, Ed came crawling out onto the spur. Just as he put up his glasses for a look, the ram got to its feet, stepped out into full view and stretched luxuriously. It was big and dark-colored with a massive set of full-curled horns broomed off at the tips.

Ed's whisper was loaded with excitement, "Big one! What will he go?"

"About forty on the curl and maybe a bit over sixteen around the base," I guessed. "That's a real trophy ram."

Then another ram appeared and another, until like magic ten of them filed out of nowhere onto a little green meadow trapped between sheer walls of solid rock and timber. They were nearly all mature males, the first two outstanding. The first one I had spotted was the most massive and likely scored the highest, but there was another almost as big that was truly classic. Both curls were fully as long but not as heavy, the horns flaring at the tips unbroomed. I marked that one down for my personal attention.

Even if the season had been open, the rams were in an impossible place. Maybe the morning would find them feeding out on the talus fans. By now the sun was dipping behind the divide, so we went to the horses and rode back to camp.

Something about sighting game and watching it changes the atmosphere of a hunt entirely. There is a suppressed air of excitement, a pleasant tension, and a keenness of looking forward to the morrow.

After supper, while a warm fire snapped and crackled in our tin stove, Ed gave me a learned dissertation on the virtues of wearing a white hat on a sheep hunt. He wound up by saying, "Eagles, rams, rabbits—anything you want—they all come to a white hat."

"You better wear it to bed," I advised. "A grizzly might sneak in and steal it. We could need its luck in the morning."

In the morning, by the time the first sun was just beginning to tint the tops of the mountains in pink and gold, we were crawling the last few feet out onto the rim of the big basin for a look.

If I live to be a hundred, that moment will always remain breathtaking. A shaft of sunlight was pouring through a notch at the top of the basin a mile to the southeast. Most of the basin was still in deep shadow, but this sun streak lit up a big talus fan like a spotlight. And on this alpine stage, sunsplashed, with its dramatic backdrop of cliffs, was a collection of game sufficient to make a man doubt his vision. Glowing like beacons in their white coats were seventeen goats feeding near the top of the fan just under the wall. A couple of hundred yards under them, twelve rams were bedded in the loose talus.

"Well I'll be—!" I heard Ed softly murmur. "The whole durn circus has turned out!"

We decided to split up and make our stalk from two directions. Ed would swing back down the ridge, climb into the bottom, and make his approach up the creek. In the meantime I would make a great circle around the head of the basin under the skyline to a strip of timber on the opposite slope. Then I would follow this down onto the basin floor. This way, if the rams moved, one of us was bound to get a shot. With a bit of luck thrown in both of us stood a good chance to get our pick of the rams.

I had a long way to go over some steep country, so I

hurried. A bit over half an hour later I reached the top of the timber strip across the basin and sat down for a breather while checking Ed and the rams with my glasses.

The sheep were still bedded down, but Ed was nowhere to be seen. Then to my vast astonishment I spotted him legging along my tracks following me around my circle. But when I looked back over the country he had planned to traverse, it was easy to see why he had given up his original plan. For about two miles the cliffs formed an almost unbroken palisade, where climbing down off the ridge top would be a very tricky business. Ed was nursing a tender ankle that had been broken in a mixup with a horse a few months earlier and showed good sense in changing his mind.

I waited while he came around the circle to cross behind me heading for another strip of heavy timber. He had not changed his plan, only his method of approach. Taking lots of time to give him a chance to get into position I eased down through the trees toward the bottom.

In two or three places the cover thinned out, forcing me to crawl flat on my belly through a screen of little trees. Once I cut a small fir and held it ahead of me as I snaked down across an open piece of slide rock. The rams continued to doze in their beds, chewing their cuds. The goats had all climbed higher on the lower shelves along the bottom of the cliff and were also lying in various attitudes of repose. They were of no interest beyond their presence, but they posed something of a problem, for they had thirty-four sharp eyes that were bound to pick me up if I bungled my approach. If the goats took alarm and high-tailed it out of there in a hurry, it would likely spook the rams before we could get into shooting range.

A look down into the basin afforded a brief glimpse of Ed's red shirt moving up a steep-timbered gully toward the foot of the cliffs. He was nearly in position and now it was my move. Taking full advantage of every bit of cover I slid into the bottom of a dry watercourse and headed directly up toward the rams.

The scant cover led me through patches of ripening grass and over the ragged flotsam of decades of snow avalanches. Once I crawled past the bleached-out skull of an old ram—relic of some tragedy seasons past. It was getting hot and sweat was running into my eyes when I finally came up behind a big boulder about four hundred and fifty yards down slope from the bighorns. Through a bit of screening herbage my glasses drew them up close. They were still bedded and oblivious to our presence, for the morning thermals were blowing down slope from the cool shade of the cliffs. But they were beyond range of my .300 Savage, a beautifully stocked, brand-new custom saddle gun built for me by the late Jack Reid, a master riflemaker famous for his craft all over western Canada. It was a very accurate weapon, my idea of the kind of rifle for carrying on a horse, but it had its limitations, requiring one to stalk fairly close.

After studying the ground in front of me for several minutes, it became obvious that it was impossible to cross it without being seen, for it was slightly concave without enough cover to hide a rock rabbit.

There was only one thing to do and that was sheer gamble. The ram I wanted was lying at the bottom edge of the bunch closest to me. The big dark one Ed wanted was at the right-hand edge of the bunch, about three o'clock in his direction. If they jumped prematurely I might lose out, but Ed was almost sure to get a chance. Deciding to try something, I took my rucksack in one hand and my rifle in the other and slowly stood up in plain view of the sheep. Instantly every ram had his eyes riveted on me and the cud chewing stopped abruptly. Very casually I stepped out, moving up the slope to my left. Not a ram moved an eyelash. Every few steps I stopped to pick up a rock or a plant, examine it, and toss it aside. Once I even turned my back on them, looking away down valley as though rams were farthest from my mind. I looked casual, but did not feel that way. Inside every nerve was singing with the excitement—set on a hair trigger as the range closed—instantly ready to flop into a comfortable shooting position and start burning powder.

Hardly daring to breathe I came to another boulder a scant two hundred yards from the sheep. Still very casually, I put my rucksack on it and sat down to lay my rifle across it. The action was about to open.

The big ram had had enough. He leaped to his feet like a goosed bronc and went barreling out across the rock slide with two lesser henchmen at his heels. The rest of the bunch jumped up and milled in indecision. I watched the big ram go and waited. Just as he reached a gully Ed's rifle thundered and the next instant the ram's feet appeared briefly as he somersaulted into the gully. My ram instantly careened off to the left through a scattered boulder field, way ahead of the rest of the bunch. Swinging my sight out a bit ahead of his nose I cut loose. He stumbled, but recovered and kept on. Again I swung with him and fired and yet again. Each time the meaty plop of a striking bullet came back to me, but the ram turned straight down slope and seemed to speed up till he was flying. At my fourth shot, he turned a complete end over end in the air and came down sliding with his horns clattering in the talus among a cloud of dust.

When I turned it was to see Ed standing over his ram and minutes later I was helping him hold a tape on those big horns. They measured a bit over forty inches on the curl and slightly over sixteen inches around the base—a Boone and Crockett trophy.

My ram also qualified for the record book, but the distinction was only a byproduct of what had been one of the most dramatic and exciting stalks of my many years' experience in the wilds. When I look at his beautiful head on my wall as I write this, it is not with a feeling of having competed successfully on a statistical scale, but with a memory of a great hunt in good company, the sound of horse bells echoing off the cliffs and the smell of wood smoke—the really important things, of which the killing is an anticlimax.

THE CHALLENGES connected with hunting as a professional guide were perhaps just as attractive as the profits gained by the business of taking people for hunts among the Rockies. For a man to be a good hunter was not enough, for it might be easy for him to stalk and take a particular ram or bull elk by himself, but it was always much more difficult for him to get a client up within good shooting range in a calm enough frame of mind to make the best use of his weapon.

There came a time when stalking and killing big game meant nothing much more to me than obtaining some needed meat. There was no challenge or much excitement any more; but acting as a guide was something else. For when dealing with a client—even the most experienced—there are always some unknown quantities to consider. As the poet Robert Burns has said, sometimes "the best laid plans of mice and men gang aft agley." There came times when experience meant very little and circumstance just took over to make things interesting, frustrating, and sometimes just plain complicated.

I remember a memorable moose hunt that progressed along the lines of a comedy and ended up about as unprofessional as anything could be.

Bud was a hunter from the deep south out on his first Canadian big game hunt by wilderness packtrain with me. Never having hunted anything bigger than a whitetail deer, he naturally succumbed to the temptation of getting a big bull moose for his budding trophy collection. Ordinarily this would have been an assignment about as easy as shooting fish in a barrel in the midst of my Flathead hunting territory, the biggest part of the problem being to find a bull with sufficiently impressive antlers to make a decent trophy. There were plenty of them to look over, especially down around Beaver Creek, where we were camped.

Everything was going well on this hunt but the weather. It was absolutely putrid—wet and miserable with snow every night and rain every day. We rarely saw the sun, and when it showed it was with a watery weakness that left no cheer. The creeks were all in flood and the Kishaneena running in front of our tents, pitched among a scattered grove of big spruces and pines, was belly deep to a horse and as big as a river.

For about three days Bud and I had been hunting moose within easy walking distance of camp up along tributary Beaver Creek, and each of those days started and progressed about the same way. My brother John was horse wrangler on that trip, and every morning he would bring in the horses from where they were pasturing up Beaver Creek valley. He would disappear up the trail shortly after daylight toward some beaver meadows and just as regularly come tearing back to report seeing a big bull moose half a mile or so away.

Bud would then grab his rifle, and we would take off afoot through the snow to get the moose, but invariably the bull had disappeared. The snow was about a foot deep, heavy and sloppy wet. The timber was weeping big wet drops that soon found every chink and crack in our rain gear. There were moose and horse tracks in a complete hodgepodge of sign for two miles or more. Inevitably we found moose, plenty of moose, but always cows, calves, and boot-jack bulls. We never even caught a glimpse of the two or three big bulls in the valley, and after hours of looking and getting wetter and wetter, we would slosh back to camp, tired, hungry, and disgusted, looking like a couple of half-drowned muskrats.

After about three days of this, Bud was feeling low. I was feeling anything but cheerful myself, but I tried to be light-hearted about the whole thing by telling him to cheer up; he would likely get a good moose from the cook-tent door some morning. He grunted at me, plodding along with his nose slanted at the ground, and it was not hard to see he thought my humor was about as flat as it could get.

Came the morning of the fourth day and as usual it was snowing hard as John headed out to get the horses. For once he did not come back to report a bull moose, and about an hour later I stepped out of the cook tent to listen for the bells. Down by the horse corral, sixty or seventy yards away through the big trees, there was what looked at first glance like a horse partially hidden by the trunk of a big spruce. But then it moved into full view, and it was a huge bull moose.

Bud's hunting partner was in the cook tent visiting with the cook. Bud was in their bed tent lying on his sleeping robe reading a book while the tin stove roared in one corner. I sneaked up behind the tent out of sight of the moose to tell him to get his rifle and come shoot a moose standing in the corral gate. He thought I was kidding and stuck his head out

the flaps in front for a look. Then he came apart at the seams
and without thinking grabbed his partner's .270 and his own
.30-06 ammunition. Both rifles were almost identical except
for the caliber, and now Bud came out all spraddle-legged,
looking for trouble as he tried to stuff the right ammunition
in the wrong gun. He fumbled with the weapon, trying to
close the bolt for several precious seconds before I realized
what was happening, grabbed the thing, and more out of
pure luck than good management, somehow managed to
extract the jammed cartridge with the nose of a small pair
of pliers that I had in a hip pocket. Slipping into the tent I
got the other rifle, loaded it, came back out, and handed it
to Bud ready to go.

That fool moose was still standing there broadside patiently waiting, about as big as the side of a barn, when Bud
threw up the rifle, sighted, and yanked the trigger. The
bullet went away harmlessly through the scenery and the
bull jumped off the bank into the river to head out across it
at a trot. Bud jacked another cartridge into the chamber and
took aim.

"For Pete's sake!" I yelled. "Don't shoot! Wait till he
comes out the other side!"

But Bud never heard me. Kerpow went the rifle, and
down went the moose, with a great splash right out in the
middle of white water as dead as the proverbial stone. At that
moment I could think of nothing to do but swear as the
flood swept the bull downstream and slammed it into a big
rock, where it hung up as the current twisted those big
antlers around, jamming them solid against the rocky bottom.

Bud broke in on my mumbling by opining in his deep
southern accent that I was the funniest man.

"Heah Ah been tramping around through wet snow up to
mah rump pockets for about a week looking for a bull moose

and getting nothing but soaking wet for mah trouble," he said somewhat plaintively. "Now Ah got one and yo' can do nothing but sweah!"

"Did you ever try to get a dead bull moose out of a place like that?" I asked him.

"Sho' 'nuff!" Bud exclaimed as he looked at what could be seen of the bull sticking up out of the white water. "Ah reckon maybe Ah should have waited till he got out the othah side. Why didn't yo' say something?"

To make a long story short, my brother came back with the horses and we saddled a couple of big strong mounts. Then we took our ropes and rode out to retrieve about fifteen hundred pounds of dead moose. Without getting wet, we managed to drag the carcass out into about six inches of water at the bottom of the bank. There we had to leave it. It was a freezing, miserable job skinning the cape and butchering the animal out, for there is nothing more awkward than a big moose in water, but we laughed and kidded Bud as we worked. A hunt that had got off to a slow start was on the move; we had enough moose meat to last the whole trip and then some, and Bud had a good trophy.

To be a successful mountain guide, one must know how to get along with people, be firm yet diplomatic. A man has to know how to laugh at himself and when to laugh at others. He has to have some sense of the value of giving service and at the same time retain the authority of command. For the people he deals with are, for the most part, accustomed to giving orders to others, yet they are out of their element in the wilds and appreciate being able to follow the direction of a man who knows. A good guide knows how to take charge in an emergency and at the same time has a keen sense of the value of showmanship. I recall times when things fell into place so smoothly it looked about impossible to go wrong, and other instances when every-thing went awry.

Bert Riggall was not only a master guide; he was also a good teacher, and sometimes his teaching was subtle to the point of being beyond notice, while at other times it could be very pointed.

When bad weather hits a hunting camp in the fall, it is snowing and blowing, the visibility is about nil, and trying to find game in the high country is like looking through cotton wool; it is not much good trying to hunt. Even if a hunter does get up to some animals, it is pure luck to get a shot at a decent trophy. Knowing this, Bert Riggall and I generally holed up where it was comfortable till the weather cleared, if time and the patience of our hunters would allow.

So it happened one fall up near the headwaters of the Oldman—away up near timberline on the north fork of Oyster Creek. Our hunters were three young fellows from Minneapolis, two brothers we will call Jones, and their brother-in-law. Bert was head guide then and I was second guide and horse wrangler for the outfit. One morning we woke to find it snowing and blowing a blizzard and Bert only needed a look out the front of the tent to know this was no day for sheep hunting. So after breakfast I took my time about rounding up the horses. Bert dug a good book out of his war bag and got set to take things easy for the day. I corralled the horses, caught a day horse, and was about to turn the rest of the bunch loose when one of the brothers came over to where I was standing in the corral gate.

"We'd like to go hunting," he said.

"So would everybody," I told him, "but this kind of weather don't trail with the idea."

"We would like to go anyway," he said. "We don't mind a little weather."

"You had best go talk it over with the boss," I said. "If you can talk him into it, I'm willing to have a crack at it."

A few minutes later Bert came up to the corral and told me to saddle up, we were going hunting, and by the way he said it I knew he was not very happy about it.

About half an hour later we were on our way, and the farther we went the worse it got. That was about the wildest day I ever spent in the mountains. It was blowing a gale straight from the Arctic and snowing hard enough up along the foot of the divide so the horses hated to face it. Never saying a word, Bert took us on a tour of the country. To say we were hunting would be a joke, for most of the time visibility was about fifty yards. Along toward mid-afternoon we were coming back across the head of a big, steeply pitched gully when the wind suddenly let up and the snow and mist lifted a bit.

We looked down and there busily digging for gophers was a tremendous grizzly bear about two hundred yards below the trail. Bert stopped his horse and pointed. Nobody moved. Our three hunters were hunched up and just about frozen to their saddles. Before anyone could decide what to do, the wind came back and the snow and fog dropped like a curtain to conceal the bear. Bert started his horse again, leading the way to camp, still not saying a word.

When we were unsaddling about evening in front of the corral, George Jones walked up to Bert and asked, "What did you expect us to do back there when we saw that bear?"

"Mr. Jones," Bert said, "when you are out hunting and you see a bear, it is customary to shoot at him!" And then he went on with the unsaddling.

I was stuffing my glove in my mouth to keep from laughing out loud as the hunters looked at each other and then stiffly headed for their tent. Up there on the mountain they were too cold to move, let alone tackle a big grizzly, and they knew it. There is no teacher like experience. Those young hunters came back for years after that, but never again did they ask anyone to go hunting in a blizzard.

But of course there are always extenuating circumstances, like the time my old friend Bart and I were storm-bound up on the head of Hidden Creek, till we were both about to come down with cabin fever or what goes for it in a tent. We were low on meat and sick of doing nothing. Bart wanted a good mule deer buck, so we went out to see if we could find one.

About mid-afternoon we were crossing an open basin up at timberline, when through the falling snow I spotted a fine buck feeding about four hundred and fifty yards up a steep bunch-grass-covered slope not far from a big lone fir tree. It was too far for a shot, so I tied the horses to the dry roots of a big deadfall out in the middle of the basin, and we began to climb. Meanwhile the cloud curtain had moved down to cover the deer and hide us. I led the way purely by dead reckoning, keeping the wind right and following a gully to a spot where we were about one hundred and fifty yards from where we had last seen the buck. Still the mist and snow hid everything. We couldn't even see the tree for a while, but then, as it often does, the storm lifted enough for us to see the buck standing on the high side of the fir. Bart was in a good comfortable position to shoot and when he squeezed the trigger, the buck slumped stone dead to lodge against the tree.

"Now how do you propose to get him down out of here?" he asked after we had cleaned the deer out. "It looks mighty steep and slippery up here for a horse."

"Nothing to it," I assured him in a moment of recklessness. "You bring your rifle and take your time. I'll see you at the bottom."

So saying I twisted the big antlers back like the handlebars of a bicycle, straddled the buck, and took off down the mountain on about ten inches of new snow. In fifty feet I was going about fifty miles an hour and in another few steps about two miles an hour short of a free fall. Never in all his life had this buck moved at such a velocity. Everything went fine till we hit something under the snow—a stump or a rock—whereupon the buck and I took off airborne in a kind of forward loop to come down with a crash that shook the bones of my ancestors in Scotland. I landed flat on my back and the buck came down beside me, the long sharp prongs of his antlers burying themselves in the ground about half a foot from my shoulder. It was about as close as I will likely ever come to being killed by a dead deer.

Aware that my foolishness was beginning to show, I compounded the folly by playing it through. Once more grasping the antlers and swinging the deer around, I straddled it and headed for low country. Sometimes I was flying and sometimes I was dipping low enough to scoop up a blinding sheet of snow. More by luck than any kind of skill I somehow contrived to miss anything solid enough to upset my going. It was a wild, crazy ride that finally came to a stop right in the middle of the basin.

There my horses looked up to see me descending out of the storm riding a buck, whereupon they promptly broke their bridle reins like string and headed for home. I just sat there watching them go as Bart came along to sit on a log. He was shaking with laughter.

"Oh boy!" he chuckled. "What a picture! I would give a good deal to have had the camera I left in camp."

He was completely good-natured about the two-mile walk back to camp, where the cook had caught our horses and tied them up.

That trophy was a good one, but I doubt if the size of it is what makes Bart grin when he looks at it hanging on the wall; for I will gamble it is not too hard for him to conjure up a picture in his mind of a wild ride down a mountain slope through a snowstorm.

No two hunters are quite alike, and by the same token no two hunts are alike either. Of course there is always the unknown of weather to contend with and the movement of game, but sometimes a guide has nobody but himself to blame for getting into some unusual predicaments.

One time I wrote a chapter for another man's book on the habits of elk and elk hunting, for which he gave me full credit. However, in that account, I put two or three experiences together to illustrate the ultimate hunt—a real picturebook kind of hunt for a big trophy bull—the kind one dreams about but rarely encounters. It was done to illustrate just how a hunt should be conducted in rough country where problems of wind and topography can be tough. I must have done a fair kind of job, for some years later a New England sportsman read the book and promptly booked a trip for himself and a friend, although at the time I had no idea that my elk story had anything to do with it.

Come early October that year, Doc and Jim showed up all primed up and enthusiastic for a big-game hunt, a new experience for both of them. We went by packtrain through the Rockies. From the beginning we were blessed with some of that wonderful golden, tangy kind of weather, when it freezes a bit at night and is clear and cool in the daytime with no wind to mention and the atmosphere so brilliantly clear that peaks fifty miles away were sharply etched against the sky. There was a bit of snow lingering on the mountain tops, leftovers from a couple of earlier storms, and the whole country was in autumn dress above timberline, with the timbered lower country all draped in red, orange, gold, and green. It was the kind of weather where one can roam through the Rockies completely satisfied, whether one shoots anything or not.

For the first part of the hunt, my head guide, Wenz Dvorak, took Doc with him while I guided Jim. For openers, both hunters stalked and killed two fine, old mountain goats —big-bearded billies with well-grown pantaloons and respectable horns. Then Doc got a chance at a handsome silver-tip grizzly and collected that after an exciting climb. Meanwhile Jim and I stalked and collected a very good mule deer buck.

A week had not gone by and we had four good trophies. It was the kind of trip where it appeared we could do no wrong, so I was a bit surprised to note that Doc was looking a bit long-faced and thoughtful one morning as we got ready for the day. He was reluctant at first to loosen up and tell me what was bothering him, but finally he began to talk.

First he made it very clear he was casting no reflections on Wenz's ability to guide and hunt, but he had been dreaming about getting a bull elk with me. As a matter of fact ever since he had read the book he had been planning it, and would I take him? Naturally I was complimented, but I was also in something of a spot; for I had written up that hunt as it should happen maybe once in a lifetime—the supreme hunt—and nobody knew better than I that stalks rarely turn out that way. Assuring Doc there was no problem, I also told him not to be too surprised if things turned out a bit different from what he expected. We might run into a monster bull a few hundred yards from camp, step off our horses, and take him: end of hunt. Or we might get into a spot where a bull elk got away leaving me looking like the world's champion amateur. But there were good odds that he'd know he had been on an elk hunt before it was over.

Doc was whistling lightheartedly as we headed up the trail that morning, and I was still wondering about guides who choose to write on the side. Doc was obviously convinced he was out hunting with the world's best elk hunter, an opinion I did not exactly share, for a lot of elk had contrived at one time and another to prove otherwise. It was a funny situation to be caught in and about all I could do was try to make the best of it.

There was a long, steep-sided, heavily timbered canyon with its upper end twisted and cut by a series of side canyons and gullies where it opened up at timberline at the base of a big mountain a few miles away. We hadn't hunted it yet, so it was undisturbed, and also a sure-fire hangout for elk. It was a tough place to get into, with a lot of heavy deadfall and underbrush to get through, but it was a piece of country well worth the effort.

After three hours of steady riding, twisting back and forth through the timber dodging logs and brush, we finally came out on the first of a series of big, open avalanche tracks, where winter accumulations of snow came down annually to keep swept clean several open lanes through heavy timber as thick as the hair on a bear.

It was noon, so we stepped down off our horses to have a look up the valley and eat lunch. I climbed up to a big upturned stump to use my binoculars while I ate, and immediately spotted elk about a mile ahead near the top end of the valley. Motioning for Doc to come up, I proceeded to give the prospect some study.

The more I looked the more elk showed up. There must

have been fifty of them—cows, calves, and bulls of all sizes up there on the broken ground at the foot of a wall of rock a thousand feet high. Behind them a bridal veil falls twisted and danced in the wind as it plunged from a high basin down into the head of the valley. Lifting in a snow-streaked mass of limestone and reddish argylite beyond, like the backdrop of a stage set for a drama of the gods, was Mount Yarrow, just under ten thousand feet at its crest. The picture framed in the field of my glasses was something to put a knot in the throat muscles. There were elk feeding and lying all over the place among green slopes laced with the red and gold of scrub willow and birch, all backed by the falls and the face of the mountain. Overhead the sky was a brilliant blue with a few puffy clouds sailing blithely on a soft southwest wind. It was a dramatic setting for a stalk and Doc was enthralled, his lunch lying forgotten beside him as he looked at it through his binoculars. But having been there before many times and suffered some shattering defeats, I knew exactly what I was up against; for besides being as grand a place to hunt as one ever sees, it could also be the trickiest. The direction of the prevailing wind, the warm sun, the steep cliff at the far end of the valley, and the twisted, broken nature of the place all contrived to make the air currents tie themselves in knots.

Elk have marvelous eyes and excellent noses. They are smart animals that take no chances when something alien shows up in the vicinity. With fifty pairs of eyes and as many keen nostrils to consider, I had my work cut out for me, for they were located in a real fortress of defense. Doc's dream stalk had a good chance of being turned into a fiasco unless I played every detail right.

I never saw anyone enjoy anything more than he did sitting there anticipating that stalk. He was as cool as ice and as confident as a man can get, but underlying this exterior it was easy to see he was bubbling with excitement and delight. Determined not to fumble if it could be avoided, I stayed on my root studying the lay of the land through my glasses. The more I looked the worse my problems seemed to be: the sight was beautiful but also formidable. It was plain as mud that up there with the elk the wind was running absolutely wild.

There was a combination of thermals and regular wind currents conflicting, so that all the normal rules failed to fit, making any ordinary kind of stalk look almost impossible. For about an hour or maybe longer I just sat and looked, wondering what to do.

By now the sun was dipping far enough to the west to indicate that time was slipping away. I thought of using my call, but this was the tail end of the rut and the bulls were not showing much interest in the cows. Even if it worked, probably only a minor bull would react. There seemed nothing to do but try a stalk. With any other hunter I would have made a play and taken my chances, but with Doc I was undecided.

Another half hour went by, when somewhere away up the mountain back of the elk the wind got hold of some mountain avens' fluff and now the seeds came sailing down, all backlit in the sun like tiny, shining snowflakes. Screwing the focus of my glasses down fine, it was possible to see these little natural parachutes for a considerable distance, and they were mapping the vagaries of the wind as accurately as anything could do. Watching carefully I got the complete picture of the air currents etched in my mind till it was memorized to the last detail.

Beckoning Doc to come, I led up toward the elk through the stringers of timber and slides. What followed was about the most twisted route anyone could dream up. I could see the questions on Doc's face as he trailed me up one draw over a saddle into another, down that in a half reverse around the end of an old glacial moraine, and then along an inclining scrub-covered bench to still another gully. Still the avens' fluff came dancing on the wind, so I could continuously check and correct our route until we finally crawled up the back of a knoll among the wreckage of some avalanched timberline trees.

When Doc poked his head over a log on top of the knoll to look, he hissed through his teeth in excitement, for lying just across a narrow canyon on an open grass-covered slope were five bulls. Two were small, one was mediocre, but the other two were fine six-pointers. One of these was a beautiful trophy, obviously the bigger of the two, and this one Doc immediately picked as his. But I whispered to him that there was still another bull, the old herdmaster, lying hidden in some brush just back of them. I had seen his antlers a couple of times, and he was a monster. But Doc demurred, telling me that the biggest of those in front of us was what he wanted. It would just fit a blank place on his trophy-room wall. When I nodded, he edged his rifle forward, but by this time I sensed a chance for a bit of fun.

Laying a restraining hand on his shoulder, I pointed to a place a few feet in front of us where he could sit and rest his arm as though shooting from a bench rest. Doc was obviously horrified at the idea of us showing ourselves, but I slipped off my rucksack and casually stood up, signing for him to follow. The bulls were looking away and half asleep. None of them flickered an ear our way as Doc slid into position, hardly venturing to take a breath and popeyed at the daring of this maneuver.

Again he got ready to shoot, but I stopped him a second

279

time. First, I carefully arranged my rucksack under his arm on top of the log in front of him. Then I whispered for him to wait till I got the bulls on their feet. By now Doc was entering into the spirit of the occasion. He was remembering that he was out with the world's foremost elk guide, and his blood pressure was somewhere closer to normal than blowing the safety valve.

So I yelled at the elk to stand up. They didn't budge an inch. Again I yelled and still those bulls lay dreamily chewing their cuds. This was fast becoming ridiculous, so I stood up in plain view waving my hat and yelling. That brought them alive and instantly they were on their feet, Doc's choice broadside at about one hundred and twenty yards.

"Okay, Doc," I said, "take him! Hold a bit low at the base of his neck just in front of his shoulder."

There was a pause and then the rifle cracked. The big bull dropped where he stood with a broken neck, clean killed so dead he never kicked.

At that moment the whole mountainside erupted elk—bulls and cows spilling out of the hollows and scrub timber all over the place, milling for a few seconds and then lining out at a long gallop down across the creek and then up along a timbered spur toward the crest of a high-flanking ridge.

Among them was a great bull, about as big as one ever sees, his mahogany-colored antlers with every ivory-tipped tine shining in the sun. As he topped a point above us, he stopped to turn his great head and look back, proud and wild—a royal stag. Then he was gone; we were alone with not another living thing in sight.

Doc stood up and said almost reverently to the world at large, "Well I'll be damned!"

I didn't say a thing. There comes a time when there is no point to redundant talk—time to leave good enough alone.

So reputations are made—taking some skill to be sure, also a measure of experience, but more of luck and a willingness to play according to the tip of the hand of fate.

As always it was the stalk that lingers in the mind, the thrill of coming up on a great animal close enough to be sure the scoring would be clean. The killing was a secondary thing—the anticlimax wherein the prize was plucked as proof of where we had been and what we did.

# *The Changing Face of the Land*

IT WAS BITTER COLD ABOUT THREE HOURS AFTER SUNDOWN when we swung the long line of steers down off the southern slope of Belly Buttes. They were strung out for half a mile or more, plodding patiently, glad to be quartering away from the bitter northeast breeze. The driving frost crystals that had been cutting our faces for two long days had stopped, but it was colder—maybe twenty below zero—and the prairie lay blanketed under a foot of new snow ghostly white in the brilliant light of the moon. Every steer had a plume of steam over his head, a sort of flag defying the elements and proving they were still alive and warm even though their hides glistened with frost. The prairie was asleep; no coyote howl broke the stillness over the frosty, creaking sound of hoofs. It was iron-cold and hostile land, yet magnificent in its sweep of immeasurable grandeur.

I contemplated it from under ice-hung eyebrows, hating it a bit right then, for I was tired and my horse was tired; we were a part of a bobtailed crew that had been working short-handed and freezing it seemed for a solid week to get this herd moved down off the summer pasture near the mountains to a feed yard a hundred miles away. Across the line of steers back toward the drag, his outline a bit hazy through the fog of their breathing, George Scout, a Blood Indian, rode hunched up and miserable. He was a pale contrast to his grandfather, his blood watered down somewhere along the ancestral backtrail when a white man followed a dark-eyed beauty into a teepee. At that moment he was near freezing in his store-bought clothes, his feet like ice in boots and overshoes and the lobes of his ears showing a dead grayish-white from under the flaps of his wool-peaked cap, frozen solid. I wondered at him not having the sense to dress like an Indian and stay warm; but then remembered that he wore his hair short like a white man. Maybe feeling he was neither one thing nor the other, he was trying to make a choice, and I was sorry.

For here was a prairie wilderness as yet unscarred by fence or plow stretching away like an ocean of snow and grass to the horizon. It was Indian country, buffalo country, wild and free as the wind, although only the shades of the buffalo walked with red men dressed in paint and feathers and soft, tanned buckskin. Thinking of this, my tired hate of it faded to nothing, bringing a grin that hurt my wind-burned lips; for here was I savoring what was left—the grass and the snow and the immensity of it—from the back of a horse, with some steers instead of buffalo, to be sure, but at least I was dressed in smoked moccasins and a buckskin jacket, beaded and fringed, that a Cree woman had fashioned many miles to the north, and as sure as I was sitting my saddle I was warmer than George.

Ahead, the long line of steers lifted and dipped gently over the soft, curved breast of the hill, so that a tiny winking light was revealed in the distance. At first I thought it was a star where horizon and sky blended together, but after watching it a while I knew it was a lighted window— probably the winter cow camp where we were headed for the night.

The loneliness of a winter night on the face of the great western prairies is a tangible thing—a thing so much a part of the country it is accepted by most; for to do otherwise is unbearable. It is then that one is aware of the immensity and realizes that man is but a speck on the face of it, which can be a devastating and even destructive thought to those not accustomed to it. To a rider, the animals he moves with are not just something by themselves but a warm link to his identity, welcome by their mere presence.

The steers seemed to know they were headed for feed and water, for they picked up their feet, and here and there along the line one bawled as though anticipating a full belly and a warm bed ground.

Then the land dipped sharply into the head of a coulee down onto a hidden flat sheltered from the wind, where the lighted windows of a cabin showed on the slope beyond by a spring. A door opened, throwing a broad shaft of light out

across the snow, as we swung the steers down off the slope
past a shed and corral toward a fenced pasture pointed out
by a man swinging a lantern.

This was the winter headquarters of Cecil Tallow's
spread, a sizable ranch operation he managed up near the
head of St. Mary's Coulee, about fifteen or twenty miles up
from the river just below Grandfather's old ranch. Cecil was
a treaty Indian with just enough white blood, he claimed, to
make him want to work at things important to white men.

He helped us look after our horses after we had shut the
gate on the cattle. The herd had plenty of good grass and
water, the horses were fed, and we had some of the cold
stamped out of our frames as we headed for his cabin. It was
a spacious affair of several rooms. His wife was smiling and
busy as she prepared supper for us.

She was a striking-looking woman, taller than her husband, with the classic features of an Indian. But her eyes
gave away part of her ancestry, for they were blue, belying
the black hair and dusky skin. She was the daughter of a
Blood woman and Dave Acres, one of the original founders
of the infamous Fort Whoop-up. Our host and hostess were
past middle age, although neither appeared burdened by the
years. They were delighted to have company and soon we
were sitting down to a feast I will never forget.

The steaks were done just right, the biggest T-bone cuts
I have ever seen; prime beef a good inch thick, they must
have been cut from a four-year-old steer carcass. With
fried onions and boiled potatoes on the side, it was a feast for
the gods and hungry cowboys. Washing it down with big,
thick mugs of smoking hot coffee, we forgot all about being
cold and tired. Even the gloomy George was grinning, with
beef fat running down his chin.

After supper we stretched our legs comfortably in front

of a heater while Cecil regaled us with tales of the cow camps and prairies, for he had worked as a cowboy on many of the big outfits as a young man. We sat back and listened raptly, for he was a grand storyteller, and one could smell the smoke of branding fires and hear the popping of saddle leather as some notorious bronc swapped ends and sunfished to unload his rider. In that warm cabin out there in the midst of the prairie vastness, we were treated to some history told with the color and grace of a master raconteur—a man who was not only good with words but also eloquent with his hands.

Finally Cecil got around to telling us how he had worked on a crew with a great bronc rider, a man who had ridden rough string for one of the big ranches in the old days. I knew him when Cecil described him, for he was my uncle for whom I was named, long dead, his bones resting in France among a host of Canadians killed in action during World War I.

When I told Cecil this he looked at me in astonishment.

"What!" he exclaimed. "But I t'ought you were mixed blood!"

There was a roar of laughter all around. Later, Jack Ecklund, the foreman of the drive, remarked on the side, "Serves you right going around dressed like a squaw man!"

Just the same, that fringed and beaded moosehide kept me warm on that ride, and the next morning as we strung out under frosty stars before daylight, I was most happy to be wearing it. That was the last day of the drive across the virgin prairie, and had I known it would be my last ride over that great sea of grass, the thought would have saddened me.

A comparatively short time later, the Indian Council under the direction of the Indian Agent and with the blessing of the department heads in Ottawa, leased the biggest part of the Reserve to white farmers. So one of the last existing pieces of real buffalo grass was plowed under, a stupidity and a waste, for the grain grown on it has only added to the mountains of surplus piled on the Canadian plains. The methods of farming have not been carefully controlled; the government does not know or does not care. The Indians could not be expected to know, for only a century ago they were hunting buffalo. But now when the warm chinook winds come roaring down out of the mountains in winter or spring, reaching velocities sometimes over a hundred miles an hour, I look out from Hawk's Nest, our home on the tip of a hill next to the mountains, to see the eastern horizon a ragged black of flying topsoil.

It is more than just a reminder of the stupid paths of so-called progress; it is a sadness, a nostalgic sadness, for without even closing my eyes I can conjure up the sight of a long line of steers walking over their knees in snow and grass with plumes of steam above their heads. To be sure they were just a first substitute for the buffalo, but the land did not die slowly in their wake, nor were the cold clear springs choked with drift dirt.

THERE IS A TWISTED CANYON not far down from the head of Drywood Creek, a place where over the thousands of years since the great ice age passed water has carved deep into the solid rock. Here and there the sides are undercut, so you can stand looking up at rock projecting over your head like shelves—a sort of terrace in reverse. Sometimes during summer heat, the mountain sheep come to drink from the stream in the bottom by following a side canyon in along airy causeways with nimble feet able to leap from one tiny ledge to another with the utmost precision. It is always cool even during the hottest summer day, for the ice-cold water leaps with abandon from ledge to ledge, throwing carefree spray and gathering it up again to go swirling and sluicing down chutes, resting here and there in pools clear as mountain air, effervescing with bubbles. In mid-canyon there is a falls fifteen or twenty feet high where the whole stream pours over a ledge like the lip of a jug into a deep, wide pool. Above this there are no fish, for trout have never been able to pass over this natural barrier.

For some time while enjoying the process of growing up that place was mine—a secret spot where I went alone most of the time, without even my brother's company, to fish for trout. It was clean and lovely along the canyon floor, with graceful ferns growing here and there out of cracks in the rock walls and islands of reflected sunlight dancing on the red and green argylite formation where it bounced up off the mirror surface of a pool. It was a favorite place for dippers to nest—the water ouzel, a cheerful, industrious, slate-gray kind of oversized wren that builds its thatch-roofed nest where the spray of falls plays over it day and night. Why dippers do not die prematurely of complications akin to rheumatism is a secret well shrouded by nature; but they thrive, and many times I was amused to watch one standing on a rock awash with ice-cold water, dipping and curtseying in a most carefree manner. And sometimes I saw one feeding completely immersed in water as it dug and scratched for bugs like a hen among the stones.

This was a pristine place, where sometimes a mink moved down over the stone steps from pool to pool like a brown silk ribbon. Once a mother and four half grown kits played and rolled happily on a tiny spit of sand near a pile of weathered driftwood not a dozen feet from my toes, completely unaware of me—proof that the boy had learned to move like other wild things with a mutual joy of living in his heart. Looking up out of the canyon, one could see mountains showing here and there, and sometimes up in a rock-framed strip of blue sky and clouds, a golden eagle swung in lazy, graceful circles.

There was a kind of magic in the place—something strong that drew me back repeatedly. So it was one hot afternoon when I first saw the great bull trout, and thereby was born what may be well branded an obsession lasting about a year.

I had been fishing for cutthroats, and when four or five fat colorful beauties were strung on a forked willow stick sunk in the water by a rock, I climbed up onto a ledge over the pool to enjoy the coolness and watch the fish. I had been there perhaps half an hour when suddenly, out of a storm of bubbles playing up from under the falls along a sunken ledge, a monster fish swam out into a patch of sunlight. The sight of it almost caused me to fall from my perch, for this was by far the biggest bull trout I had ever seen. Hung there in the current with its great ivory-trimmed fins set to hold it steady, it looked like some kind of pastel-colored submarine.

A really big trout does not get that way by being stupid. Almost always the angling for such a fish is a long drawn-out battle of wits—an exchange of guile and intelligence, a sort of piscine reaction to the exigencies of the moment on the part of the hunted, while the hunter tries everything he knows and then invents some approaches he never heard of before; and all the while he may be seeing monstrous fish in his sleep even when miles from the trout's lair.

So it was when I found the big fish in the falls pool. Despite various sneaky approaches at all hours of the day with a vast variety of lures, that one either found a weak spot in my tackle and broke loose, or worse yet, just ignored my offerings with cold, fishy disdain.

All summer I worked to catch that fish, and when fall came around with its great clear ice frozen on the slack places along the edges of the creek, the season for fishing was gone and the big bull trout had moved downstream to a suitable wintering hole somewhere in the deep water of the river.

During winter, when the creek was trapped in iron bands of ice and the hills and mountains were blanketed deep in snow, I sometimes dreamed up a vision of that great deep-green fish lying on the bottom of his favorite pool. It was then that I sometimes got reprimanded by Mother or the teacher for daydreaming when I should have been listening.

Winter crawled by interminably, but was finally warmed by the spring sun to go into retreat. The creek shook itself free of its shackles, roaring and frothing, full of forest debris and silty washings from its first wild rush on the long journey to the sea.

When the water cleared and lowered enough for me to look into the big pool at the foot of the falls, the big trout could be seen occasionally through the slick window of a boil in the current, finning in the water's rush beside the overhanging ledge covering his lair.

The sight of the monster was akin to some kind of medicine, which made my heart pound and my hands shake with excitement while readying tackle for the reopening of the contest. It was a heady stimulant, making my blood sing and the feel of tension thick, yet so intangible it was bound with hope well watered down with the knowledge of possible defeat. It was the pure excitement of trying to catch this fish, knowing that at any time my hook was in his element my arms could feel the sudden jar of a strike—the telegraphing of fast action coming up the line, electric in its potency and wonderful to feel. It was a kind of primitive contest between a boy and a great fish, wherein the fish had its life at stake but the boy had nothing to lose except perhaps pangs of frustration at times. The fight to him was a whetstone, sharpening and honing a determination that was to become a part of him. The fish did not know it and neither did he, but that long, drawn-out contest was building the kind of character it takes to live in the wilds—a certain patience, endurance, and solid persistence that were to stand by him well in dealing with problems his life had in store. There would come times when he would be justifiably cursed by those who worked with him for being stubborn and unyielding in the face of what appeared to be certain defeat, but also occasions when those same people blessed him for it. So destiny decreed that a boy and a monster trout got their lives entwined, and when the game was finally done there was a touch of sadness in its conclusion.

For a very long time in the life of a boy my preoccupation with that fish bound me up in every spare moment, and for most of the second summer my success was no better than the first. Then one hot August afternoon while gathering windrows of hay with a team on a bullrake, the sight of a mouse scrambling for safety woke cruel genius. It was a bear mouse, a short-tailed, red-backed vole that came running from under my load. It suddenly found itself scooped up in a gloved hand and placed in the darkness of a toolbox. In due course the mouse was transferred to a tobacco tin stuffed with dry grass and with nail holes in its lid for ventilation.

Early next morning I was on my horse heading up the valley at a gallop. Upon arrival at the falls pool I was not long in tying up the horse and climbing down into the canyon, where my tackle was readied for action.

A brand-new, well-soaked gut leader, heavy and strong, was tied to the end of my line. On its tip was fastened a big, sharp, new hook sufficient for the job at hand with some strength to spare. And to this was harnessed by use of a couple of light elastic bands, the live mouse. If voles were as big as grizzlies they would likely rule the world, for they are fierce creatures in the wilds with little sense of proportion in a fight and very little fear of anything. By the time the hook was rigged to my satisfaction, I was severely bitten twice and blood was running freely from my right hand.

The sun was a warm red on the top of the mountains overhead when I took up a position on a ledge just over the bull trout's hide to cast the mouse out into the pool on a short line. Hitting the water, the little animal swam furiously for shore dragging line and leader behind it, stirring up the surface with its tiny agitated feet. Then up out of the green-blue depths, like a cutthroat rising for a mayfly, the bull trout came in a gargantuan rush that split the pool wide open and engulfed the luckless mouse like a tiny Jonah disappearing down into the belly of the whale. The surging jar of it almost jerked the rod from my hands and came within a whisker of overbalancing me into the deep water. But I hung on desperately, my tackle somehow surviving my inept, almost demoralized reaction to the strike, and the battle was on with action to spare.

At first the great fish just hugged the bottom in mid-pool, swinging its head from side to side with its mouth wide open, obviously taken by surprise and trying to shake the hook loose. Then it shot up under the falls in the white water still running deep, and I could feel the power and savagery telegraphing up the line. From there it made a slamming rush downstream—a blind, angry kind of rush that ended up in water so shallow the big fish's back was awash. Throwing water higher than my head, it then swung back into deep water in front of me.

Such a fish rarely breaks water; they prefer to fight deep, taking advantage of any kind of cover available—logs, roots, sunken ledges or rocks, which are an easy lever to break a line or leader. This one had broken my leader on two previous occasions by sawing it off on a sharp-edged ledge that was the roof of his favorite hide, and now I was threatened with the same tactics. To give him as little advantage as possible, I left the ledge, backed downstream and then waded out toward the middle of the pool in the shallows near its lower end, where I stood waist deep in the icy water with a much better angle of pull to avoid the ledge. With the rod butt braced against my belly, rearing back like a big-game marlin fisherman, I fought that trout as he surged and ran up and down the pool. There was nothing very delicate or finished about my technique—I was matching savagery and power with the same kind of action—a bare-toothed fight where no quarter was expected and none asked.

Finally the pressure of the rod, aided somewhat by the current, began to tell, for the trout's rushes slowed up and became noticeably weaker. With the rod held high, almost bent double, I backed slowly out of the pool and along a small gravel bar. He tried to turn away in one last rush as the water shallowed under him, but I bent his head away so the power of it threw him high and dry. Without wasting a fraction of a second I threw the rod aside and leaped with a fist-sized rock swinging in my hand. One solid smash and the trout lay quivering between my knees—a heavy, shining, highly colored, hook-jawed male—suddenly lifeless. For a long, trembling, breath-catching moment I stood looking down at my prize, exaltation fading to regret. I looked at the pool with the falls plunging into it, knowing it would never be quite the same place ever again and tasting the bitterness of the knowing. Picking up the fish, as long as my leg and deep through the flank, I carried it to the water to wash the clinging dry sand from the shining skin and scales already fading, knowing that I would never know the thrill of seeing him swimming wild and free in the pool again. If somehow I could have put him back alive right then, it would have been done; but it was too late. Slipping a forked stick through his gill cover, I headed for my horse and breakfast.

This day just forty years later, a mere split second in the face of time that has seen glaciers melt, no big bull trout lies finning lazily in the cold, clear current of Drywood Creek—not under the falls or anywhere else along that once marvelous trout stream. For the stream is dead, not just dead but stinking—putrid with poisonous effluents defecated into it by great plants manufacturing sulphur and other things.

Few fish of any kind live there now, and even when one is caught it cannot be eaten, for the smell of it frying would make a hungry black bear gag in disgust and turn away. Clear down from the Rockies—the still shining mountains—downstream to where the great Saskatchewan rolls in lazy turns and bends through folded hills, the whole river system is foul with waste.

The sons of the pioneers, full of zeal for profits and so-called progress, the self-righteous, blind members of local chambers of commerce, have contrived to turn a beautiful river into an open sewer with the doubtful distinction of being one of the dirtiest, most lifeless streams in the world.

Sulphur plants, sugar factories, feed lots, packing plants, and a multitude of waste from cities and towns, along with various uncounted contributions from sources as yet unrecognized, have changed the silver rope of adventure that was a clean river short years ago into a place where a boy or girl cannot safely take a drink, catch a fish, or even swim. Indeed, not very many miles downstream from where I learned to swim on the St. Mary's there have been signs along the filthy river bars warning children against wading.

The smell of the waste and its disgrace are reaching many noses. The politicians, gutless creatures they have proved to be, now sensing a change from public apathy to anger and disgust, are beginning to make clacking sounds of disapproval, like aged women remonstrating against the sins they have grown too old to enjoy. And well they might, for government is in many ways directly responsible.

Contemplate the incongruousness of the government of Canada, pledged solemnly to preserve the beauty and the wilderness of the Rockies for the people for all time in what is designated as Waterton National Park, yet allowing sewage from the various installations of that place to pour untreated directly into the river, a tributary of the Saskatchewan.

Consider, if you will, the plight of a man who was born and lived all his life along the Drywood before he was suddenly stricken in his prime and took up a chronic's bed in the hospital; a once powerful, vigorous man, a great rider who loved to train horses; a sometimes arrogant man when angry, yet wonderfully gentle with those he loved. His neighbors liked and respected him and still do, and they were understandably horrified when he suddenly fell ill from lead poisoning, lead accumulated from a spring polluted from residue from the petroleum industry running rampant and unchecked in our backyards. We know the hot flush of hard anger—helpless anger without adequate support from those in authority; even the law courts are loath to move. Our neighbor lies paralyzed, hopeless, nothing much more than a vegetable compared to what he used to be, while his wife and family consider with sad bitterness a government that has failed to recognize the problem in time or do much of anything about it. Meanwhile the giants responsible for the poisoning of springs grow fat.

Like Aldo Leopold, I am glad I am not a boy again with no wild place in which to grow up. One does not begrudge that part of the wilds now being used properly to feed and accommodate the people who have come. It is the waste that is the sadness—the blind stupidity of throwing something away that would be far more useful and valuable left in its natural state, or as close to it as possible, so that young people in their quest of adventure and their search for answers to questions in their hearts would know how things were in the beginning. It might be thought by some that the man born blind does not grieve over his handicap too much for he knows not what he misses; but this could only be true if he lived entirely alone with no means of comparison or communication with other men. Today's youth stand in judgment of their elders, some determined to change a system that has proved faulty, others attempting to soothe their dread of change by inducing hallucinations. The latter, though they do not know it, have slipped into the easier pattern of waste.

Only the snow-capped mountains standing up against the sky in front of my door show no change. They are a kind of reminder that there is yet the opportunity to repair some shameful scars on the face of this plundered land, but not as much time as some may like to think. For the hourglass is tipped, the sands of time run out, and what looks like the blue, placid waters of a paradise in the distance may be only a mirage—lifeless, sterile, hopeless, and very cold.

NOTE ABOUT THE AUTHOR

*Andy Russell was born in 1915 in Lethbridge, Alberta, Canada, and describes his education as "limited formal education, considerable Rocky Mountain variety." Both contributed to the success of his book* Grizzly Country, *acclaimed by critics and naturalists alike as a major contribution to the study of the great bear. He has spent his entire adult life as trapper, broncbuster, hunter, professional guide and outfitter, and rancher. Mr. Russell now devotes that time not occupied with cattle raising to wildlife photography and writing. His articles have appeared in* Natural History *and in a number of the outdoor magazines. Mr. Russell is married to the former Anne Kathleen Riggall and lives on a ranch in Alberta.*